Payam Dehdari

Measuring the Impact of Lean Techniques on Performance Indicators in Logistics Operations

WISSENSCHAFTLICHE BERICHTE

Institut für Fördertechnik und Logistiksysteme
am Karlsruher Institut für Technologie (KIT)

BAND 80

Measuring the Impact of Lean Techniques on Performance Indicators in Logistics Operations

by
Payam Dehdari

Dissertation, Karlsruher Institut für Technologie (KIT)
Fakultät für Maschinenbau, 2013
Referenten: Prof. Dr.-Ing. K. Furmans, Prof. Dr.-Ing. B. Deml

Impressum

Scientific
Publishing

Karlsruher Institut für Technologie (KIT)
KIT Scientific Publishing
Straße am Forum 2
D-76131 Karlsruhe

KIT Scientific Publishing is a registered trademark of Karlsruhe
Institute of Technology. Reprint using the book cover is not allowed.

www.ksp.kit.edu

Print on Demand 2013

ISSN 0171-2772
ISBN 978-3-7315-0096-4

Measuring the Impact of Lean Techniques on Performance Indicators in Logistics Operations

Zur Erlangung des akademischen Grades eines

Doktors der Ingenieurwissenschaften

der Fakultät für Maschinenbau

des Karlsruher Instituts für Technologie (KIT)

genehmigte

Dissertation

von

Dipl.-Wirt.-Ing. Payam Dehdari

Tag der mündlichen Prüfung: 11.06.2013
Hauptreferent: Prof. Dr.-Ing. K. Furmans
Koreferentin: Prof. Dr.-Ing. B. Deml

Vorwort

Die vorliegende Arbeit entstand während meiner Tätigkeit als Projektleiter in der Zentralstelle Logistik bei der Robert Bosch GmbH. Ich möchte mich an dieser Stelle bei allen Personen bedanken, die zum Gelingen dieser Arbeit beigetragen haben.

Herrn Prof. Dr.-Ing. Kai Furmans, Leiter des Instituts für Fördertechnik und Logistiksysteme, gilt mein besonderer Dank für die Übernahme des Hauptreferats sowie für die Unterstützung meiner Tätigkeit als externer Doktorand. Der Begriff "scharfes Nachdenken" hat erst durch seine Fragen an Bedeutung für mich gewonnen und somit eine erfolgreiche Promotion ermöglicht.

Frau Prof. Dr.-Ing. Barbara Deml danke ich für die Übernahme des Korreferats und Herrn Prof. Dr. Robert Stieglitz danke ich für die Übernahme des Prüfungsvorsitzes.

Mein herzlicher Dank gilt den Hunderten von beteiligten Personen der Studie Warehouse Excellence. Ohne deren harte Arbeit in den Logistikzentren wäre die Basis dieser Arbeit nie gegeben gewesen. Außerdem möchte ich insbesondere Herrn Dr. Karl Nowak als Mentor von Warehouse Excellence danken. Bei Herrn Dr. Wlcek bedanke ich mich ausdrücklich für seine Rolle als Treiber des Projektes und für das mir entgegen gebrachte große Vertrauen. Herrn Dr.-Ing. Beha bin ich für sein Coaching dankbar. Erst dadurch füllte sich das Wort Lean für mich mit Inhalt. Melanie Schwab und Dr.-Ing. Christian Huber bin ich für die endlosen Diskussionen und die Korrektur des Manuskripts dankbar. Verena Pluhar, Stefan Keiber, Patrik Spalt, Martin Kreuser, Darian Achenbach, Marius Oehms, Laura Aulmann, David Geissler und Alexander Lingen, die Ihre Abschlussarbeit im Rahmen von Warehouse Excellence geschrieben oder als wissenschaftliche Hilfskraft tätig waren, gilt mein besonderer Dank.

Ohne deren Arbeit wäre Warehouse Excellence nicht das, was es heute ist.

Mein tiefster Dank gilt meiner Familie, die in dieser Welt verteilt ist und ohne deren Liebe all das nicht möglich wäre.

Zuletzt möchte ich mich aufrichtig bei allen meinen Freunden, Bekannten und meiner Familie für die letzte Jahre entschuldigen, wo es oft hieß: *"Sorry, ich kann nicht mit ich muss an meiner Diss arbeiten."* Das hat mich am meisten geprägt, denn ich hatte vergessen, was wichtig ist.

Stuttgart, Mai 2013 Payam Dehdari

Kurzfassung

Payam Dehdari

Messung des Einflusses von Lean Techniken auf Leistungskennzahlen in Logistikzentren

Die Wurzeln von Lean Techniken reichen über 50 Jahre zurück und befinden sich in den Produktionssystemen der japanischen Automobilindustrie. Mehrere umfassende und tiefgreifende Studien bestätigen den positiven Einfluss von Lean Techniken auf Leistungskennzahlen im Produktionsumfeld. Studien im Lagerumfeld beleuchten den Zusammenhang hingegen nur unzureichend. Somit besteht zurzeit eine Lücke zwischen den Erkenntnisstand des Einfluss von Lean Techniken auf Leistungskennzahlen im Produktionsumfeld verglichen zum Lagerumfeld.

Das Ziel dieser Arbeit ist es dazu beizutragen, die erwähnte Lücke zu schließen. Dies soll Entscheider dazu motivieren und dabei unterstützen, Lean Techniken im Lagerumfeld zu etablieren.

Damit dies erreicht wird, wurde vom Jahresende 2010 bis zum Jahresanfang 2012 eine Studie mit 16 Lägern in einer Beobachtungsgruppe und 56 Lägern in einer Kontrollgruppe durchgeführt. Ein intensives Befähigungsprogram sicherte ab, dass die Beobachtungsgruppe Lean Techniken in ihrer Führungskultur, kontinuierlichen Verbesserungsarbeit und operativen Prozessen etablierte.

Die Qualität der Umsetzung wurde mit Hilfe eines Lean Lagerassessments, das im Rahmen dieser Arbeit entwickelt wurde, bewertet. Dieses Lean Lagerassessment basiert auf einer neuen Generation von Assessments, die im Produktionsumfeld genutzt werden. Der Lean Reifegrad aller teilnehmenden Läger wurde anhand des Lean Lagerassessments vor und nach dem Projekt aufgenommen. Zusätzlich wurden Leistungskennzahlen vom Jahresanfang 2010 bis zum Jahresende 2011 ermittelt.

Die Messergebnisse des Lean Reifegrads und die erhobenen Leistungskennzahlen wurden mithilfe deskriptiver Statistik verglichen und mit nichtparametrischen zwei Stichprobentests analysiert. Das Ergebnis war eine hohe signifikante positive Entwicklung der Leistungskennzahlen und des Lean Reifegrads der Beobachtungsgruppe. Daraus wird abgeleitet, dass der Lean Reifegrad eine positive Wirkung auf Leistungskennzahlen hat. Ein genauer mathematischer Zusammenhang konnte nicht ermittelt werden. Weiterhin wurde beobachtet, dass die Beobachtungsgruppe eine im Vergleich zur Kontrollgruppe stärkere Entwicklung des Lean Reifegrads und der Leistungskennzahlen aufweist.

Dieses Ergebnis trägt dazu bei, die Lücke zwischen dem Erkenntnisstand über die Wirkung von Lean Techniken auf Leistungskennzahlen im Produktionsumfeld zum Lagerumfeld zu schließen. Entscheider sind dadurch aufgefordert, sich auf die Etablierung von Lean Techniken im Lagerumfeld zu konzentrieren. Denn eine entscheidende positive Entwicklung des Lean Reifegrads, die sich positiv auf Leistungskennzahlen wirkt, ist in einem Zeitraum von einem Jahr möglich.

Abstract

Payam Dehdari

Measuring the Impact of Lean Techniques on Performance Indicators in Logistics Operations

The roots of lean techniques date back 50 years to the production systems of the Japanese automotive production industry. Several in-depth studies have verified the positive impact of lean techniques on performance indicators in production environments. Studies performed on warehouse environments have only partially confirmed this. Up until now, there has been more evidence supporting the positive impact of lean techniques on performance indicators in production environments than in warehouse environments.

The purpose of this thesis is to help close the gap between the disparities in the level of evidence mentioned above. Closing this gap should cause decision makers to support the implementation of lean techniques in the warehouse environment. To this end, a study was conducted from the end of 2010 until the beginning of 2012 that included 16 warehouses in an observation group and 56 warehouses in a control group. An intensive empowerment program ensured that the observation group established the lean philosophy in their leadership, continuous improvement work, and operational processes.

Lean maturity measurements were carried out using a lean warehouse assessment tool that was developed for this study. The lean warehouse assessment tool is based on a new generation of assessments that are used in the production environment. Each participating warehouse was measured before and at the end of the project as part of the assessment. In addition to this, performance indicators were measured from the beginning of 2010 until the end of 2011.

The lean maturity results and the performance indicators were compared using descriptive statistics and analyzed using two sample non-parametric hypothesis tests. The result was a highly significant positive development of the productivity performance indicators and the lean maturity level within the observation group. This indicates that the positive lean maturity development had an impact on the performance indicators. Further research and analysis was done to determine if a higher lean maturity resulted in a higher performance development. The result was that a positive relation between higher lean maturity and better developed performance indicators could be determined. A functional relation between the lean maturity and performance indicators could not be established. Finally, the observation group showed better results in the lean maturity and performance indicators compared to the control group.

These results help close the gap in the evidence and encourage decision makers to concentrate on lean activities within logistics operations. A major lean maturity development that results in a positive high performance indicator development is possible within the span of a year in the warehouse environment.

Contents

Kurzfassung   **iii**

Abstract   **v**

1 Introduction   **1**
   1.1 Problem Description 2
   1.2 Hypotheses 4
   1.3 Organization of the Thesis 9

2 The Lean Philosophy in the Warehouse Environment   **13**
   2.1 Genesis of Lean 13
   2.2 Warehousing at a Glance 15
   2.3 Warehouse versus Production Environment 16
   2.4 Lean Warehousing: Transferring Lean Production into the Warehouse 19

3 Literature Review: Measuring the Impact of Lean   **21**
   3.1 Measuring the Impact of Lean on Production 21
   3.2 Measuring the Impact of Lean on Warehousing . . . 22
   3.3 Tools for Measuring Lean Warehousing 25
     3.3.1 Lean Warehousing Maturity Assessments . . 25
     3.3.2 Lean Warehousing Performance Indicators . 29
   3.4 Conclusion of the Literature Review 32

4 Bosch Logistics Warehouse Assessment   **35**
   4.1 Development of the Bosch Logistics Warehouse Assessment 35

Contents

4.2 Structure of the Bosch Logistic Warehouse Assessment 36

4.3 Intermediate Result: Measuring Systematic 43

5 Design of the Experiment **47**

5.1 Warehouse Excellence Group - the Observation Sample 47

    5.1.1 The Warehouse Excellence Project - Lean Empowerment 48

5.2 Control Group 55

5.3 Method for Testing the Hypotheses 56

6 Analyzing the Lean Impact **61**

6.1 Statistical Background 61

    6.1.1 Choosing the Goodness of Fit Test 62

    6.1.2 Choosing the Non-Parametric Test 63

6.2 Analysis of Lean Maturity Development 65

    6.2.1 Lean Maturity Development of the Warehouse Excellence Group 65

    6.2.2 The Warehouse Excellence Group versus the Control Group 88

    6.2.3 Intermediate Result: Lean Improvement . . . 101

6.3 Analyzing the Impact on Productivity 102

    6.3.1 Productivity Development of the Warehouse Excellence Group 103

    6.3.2 Productivity Development of the Warehouse Excellence Group versus Control Group . . . 109

    6.3.3 Intermediate Result: Productivity Improvement 111

6.4 Review the Hypotheses 112

    6.4.1 Review of Hypothesis I 114

    6.4.2 Review of Hypothesis II 114

    6.4.3 Review of Hypothesis III 115

    6.4.4 Review of Hypothesis IV 115

7 Summary & Conclusion **117**

Bibliography   **127**

A Appendix - Warehouse Excellence Group Data Sheet   **137**

B Appendix - Control Group Data Sheet   **139**

C Appendix - Assessment Questionaire   **141**

D Appendix - Warehouse Excellence Group Assessment Results   **157**

E Appendix - Warehouse Excellence Group KPR   **161**

F Appendix - Control Group Assessment Results   **165**

G Appendix - Control Group KPR   **171**

H Appendix - Warehouse Excellence Projects Overview   **175**

1 Introduction

Karl Popper was one of the most important philosophers of the 20th century (Dykes, 1999, p.1). He believed that whenever a theory appears to be the only possible solution to a problem, people have to take this as a sign that the theory has not been understood or that the problem was never intended to be solved (Lass, 1984, p.XIV).

At the beginning of the 20th century, many managers believed that the theory of mass production was the only efficient method for production (Huber, 2011, p.1). Taiichi Ohno saw the sign that Popper mentioned and questioned the efficiency of the theory of mass production. Working during the time of the tough economic challenges that the Japanese industry faced after World War ll, Ohno believed that it was possible to surpass the conventional style of mass production and produce value for the customer with less waste and with higher efficiency (Ohno, 1988, p.2). Motivated by his belief, Ohno developed the Toyota Production System (TPS) (Ohno, 1988; Liker, 2004, p.4).

In the second phase of the International Motor Vehicle Program (IMVP), scientists benchmarked the Toyota Production System (Holweg, 2007) with the mass production methods of other automotive companies. Within the scope of this detailed study, they analyzed the effectiveness of the TPS and determined the superiority of the TPS over traditional production systems (Womack, 2007). During the IMVP, John Krafcik coined the term "Lean Production" (Cusumano, 1994) to describe the philosophy behind the TPS. Parts of this lean philosophy were transferred to other functional plant areas and industrial sectors. Terms such as "Lean Management" and "Lean Administration" also came into being (Bell and Orzen, 2011; Zidel, 2006; Pfeiffer and Weiß, 1994).

Elements of the lean philosophy also eventually found their way into the warehouse environment (Augustin, 2009; Dehdari et al., 2011; Spee and Beuth, 2012; Furmans and Wlcek, 2012). The majority of reports that analyze lean techniques within the warehouse are based on pilot studies with a low sample size or even single pilot project experience reports.

1.1 Problem Description

In-depth studies that analyze the impact of lean techniques in production environments are usually based on a combination of three evaluation techniques:

- Measurement of the lean maturity
- Measurement of the performance indicators
- Comparison of the samples with each other

In addition to five other research areas, the IMVP analyzed the maturity of the production systems within plants. The scientists also analyzed the development of major performance indicators. These performance indicators either focused on one research area or on several overarching research areas. The last phase of the research involved a comparison of the data between plants with the TPS and plants with traditional production systems. The results of the comparison showed that the sample with the TPS had superior performance indicators.

With these results, the IMVP asserted that the comparison demonstrated a positive impact of the TPS on major performance indicators. Other studies have backed up the positive impact of lean techniques on performance indicators in the production environment (Bidgoli, 2004; Hofer et al., 2012; Fullerton et al., 2003; Oeltjenbruns, 2000; Liker, 2004).

This also means that the comparison of lean maturity results (independent variables) with the development of performance indicators (dependent variables) between different samples is a verified method for proving the superiority of the TPS in the production environment.

As stated earlier, lean techniques have also found their way into the warehouse environment but the conditions of the warehouse environment differ from those in the production environment. One difference is that the less technical nature of the warehouse allows more options for process changes. Another difference is that the higher degree of manual work in a warehouse causes larger fluctuations in cycle times. Further differences are highlighted in Dehdari and Schwab (2012). Therefore, the question is raised if the lean techniques that were developed for the production environment are applicable in the warehouse environment.

However, some of the verified measurement techniques used in the production environment are also used in studies in the warehouse environment. Most studies, such as Reuter (2009), use a major performance indicator to measure the impact of the lean techniques. The Reuter study, and other similar studies that will be discussed in chapter 3, does not include a measurement of the lean maturity of the operation that improves the performance indicator (Reuter, 2009). Other studies, such as Sobanski and Mahfouz, used an assessment to analyze the lean maturity but did not relate it to the development of the performance indicator (Sobanski, 2009; Mahfouz, 2011).

All of the known in-depth, verified, and reliable studies that measure the impact of lean techniques on the production environment were performed using a combination of the above-mentioned evaluation techniques. The known studies on the warehouse environment measure performance indicators but they do not compare their results with the results of a measurement of the lean maturity or they do not compare different samples with each other. Some other studies measure the lean maturity but do not link it with the development of the performance indicators. A comparison of warehouse performance indicators for warehouses that improve the lean maturity with warehouses that use other techniques is not known. This means that the applicability of verified measurement techniques for the impact of lean techniques in warehouses is still being studied. Thus, this thesis will help to close the gap between the disparities in the level of evidence for the impact of lean techniques on performance indicators within the production environment compared with the warehouse environment. This thesis will also evaluate the applicability of using verified measuring methods in production for the warehouse environment.

I hope that by contributing to closing the gap in the levels of evidence, I can help companies make the decision to invest in resources for establishing lean techniques within the warehouse environment. This should result in benefits for them because the lean maturity in warehouses today is low and the high potential for improvement is known (Furmans and Wlcek, 2012).

1.2 Hypotheses

The research presented in this thesis is based on four hypotheses which help to close the gap in the level of evidence for the impact of lean techniques on performance indicators. The hypotheses are described below. A coordinate system was used to show the difference between the hypotheses. An indicator for the lean maturity and an indicator for the performance development were used to test a hypothesis. The independent variable lean maturity is located on the abscissa. The dependent variable performance development is located on the ordinate. It also has to be noticed that the expectation level for a clear relation between the lean maturity and performance indicator rises from hypothesis I till hypothesis IV. This hypothesis lead us to discover the relationship between the lean maturity and performance indicators step by step .

Hypothesis I: Lean has a positive impact on performance indicators but we do not expect to know if a higher level of lean maturity has a higher influence on performance indicators.

Hypothesis I is shown in figure 1.1. To test Hypothesis I, warehouses that improved their lean maturity were analyzed. For Hypothesis I to be true, no warehouse that shows a positive lean maturity could show a negative development of performance indicators and there can be no evidence that a higher lean maturity also implies a higher positive impact on the development of the performance indicators. In the example shown in figure 1.1, WH 2 has a lower performance development with a higher lean maturity development than WH 1. In conclusion, I restrict myself to show a positive impact without the necessary evidence to show higher lean maturity leads to higher performance indicator improvement.

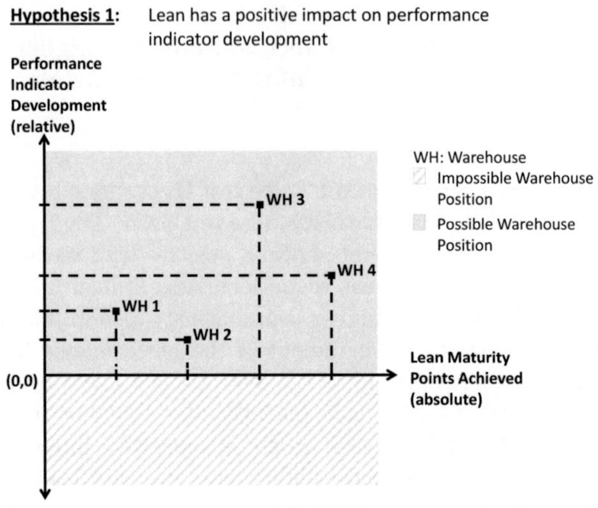

Figure 1.1: Hypothesis I

Hypothesis II: A higher level of lean maturity has a more positive impact on performance indicators but we do not know if this relation follows a mathematical function.

Hypothesis II is shown in figure 1.2. The warehouses that improved their lean maturity also showed a positive development in their performance indicators. This is similar to Hypothesis I. The difference to Hypothesis I is in the level of development of the performance indicators. If a warehouse reached a higher lean maturity level than another warehouse, it has at least the same level of performance indicator development or even higher. In the example shown in figure 1.2, the warehouses with a higher maturity level also have higher performance indicator levels. At this moment, it is not clear if a mathematical function can describe the relation between the level of lean maturity and the level of the performance indicators.

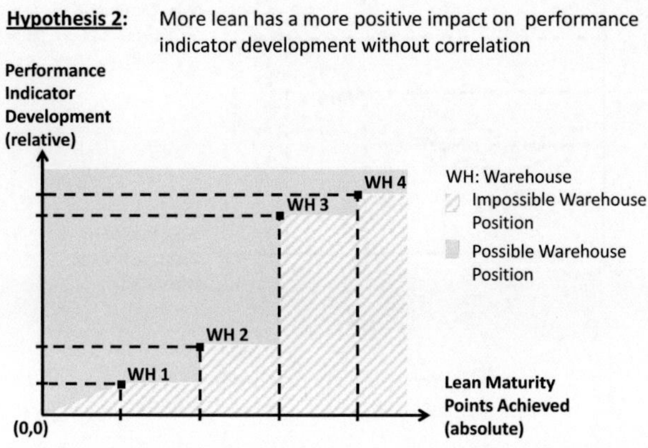

Figure 1.2: Hypothesis II

Hypothesis III: There is a mathematical correlation between the level of lean maturity and the performance indicators. A mathematical function can describe this correlation.

Hypothesis III is shown in figure 1.3. There is a clear dependency between the level of lean maturity and the performance indicators and a mathematical function describes this correlation. This function could be a straight line or a decreasing or increasing curve.

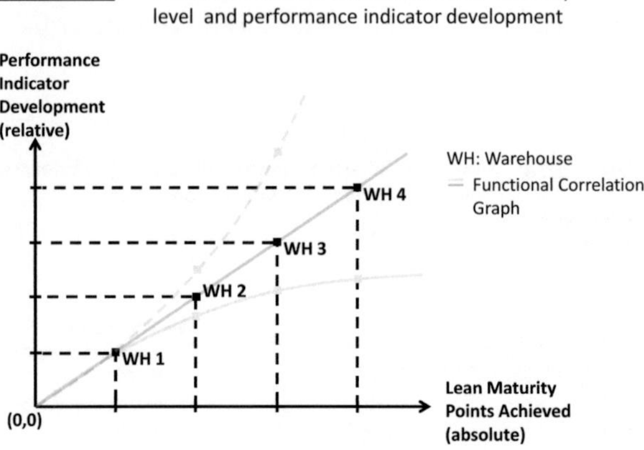

Figure 1.3: Hypothesis III

Hypothesis IV: Lean techniques have a higher positive impact on performance indicators than other approaches.

Hypothesis IV is shown in figure 1.4. The assumption in Hypothesis IV is that a group of warehouses that focuses on and improves their lean activities have a higher positive performance indicator development than warehouses that use other approaches or anything at all, instead of a focused lean development program.

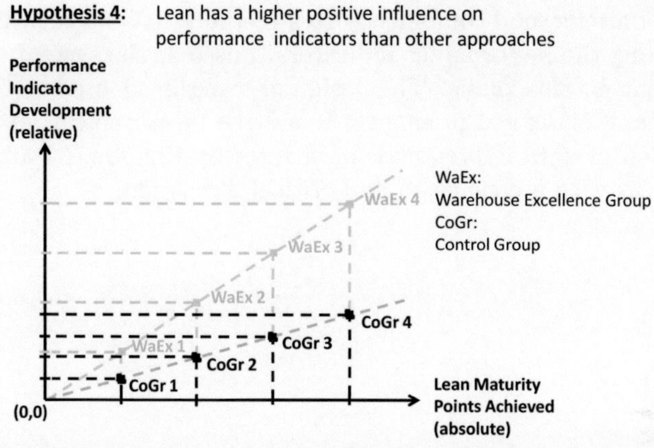

Figure 1.4: Hypothesis IV

1.3 Organization of the Thesis

Figure 1.5 shows the structure of this thesis. In chapter 1, the motivation behind the thesis is explained. To measure the impact of lean techniques on performance indicators within the warehouse environment, the terms lean and warehousing need to be defined. These terms are discussed in chapter 2. Chapter 2 also defines lean warehousing and discusses what will be measured in the following chapters.

To build on existing measuring methods, a literature review in chapter 3 identifies the relevant publications on measuring lean techniques within the production and warehouse environments. Based on this review, chapter 3 also discusses the existing methods for measuring lean maturity and performance indicators. Since a suitable method for measuring lean maturity could not be identified, chapter 4 describes the development of an appropriate method. This new appropriate method combined with a method that was identified for measuring the performance indicators is used as the system of measurement for this thesis. The design of experiment used to test the hypotheses is defined in chapter 5 and the measurement and interpretation of data is presented in chapter 6. Chapter 7 rounds out the thesis with a conclusion and critical discussion.

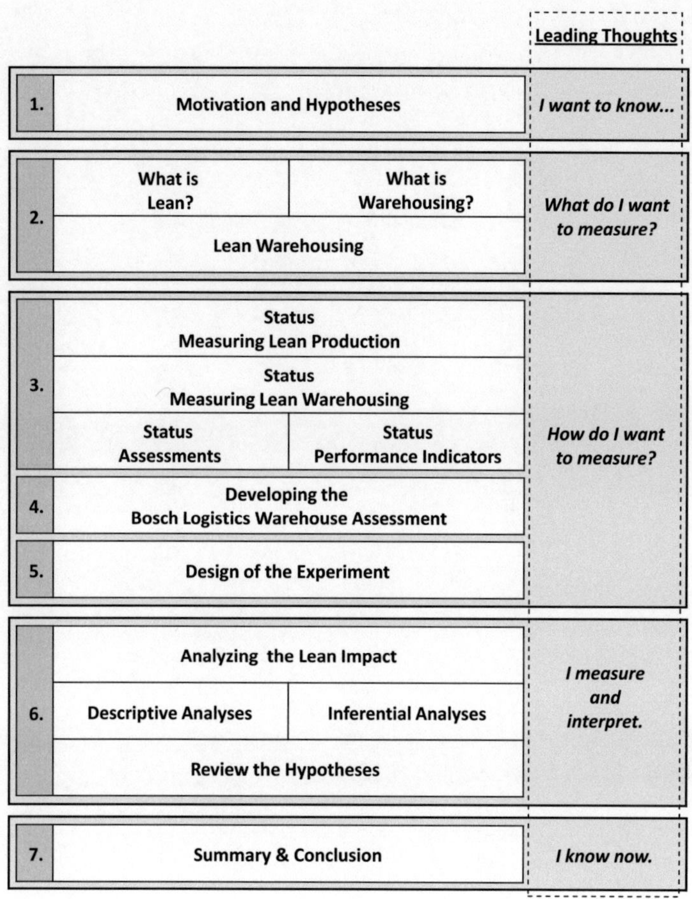

Figure 1.5: Structure of the thesis

2 The Lean Philosophy in the Warehouse Environment

This chapter first highlights the roots of the lean approach and presents the related milestone literature (see section 2.1). To transfer the lean approach to the warehouse environment, it is necessary to understand the definition, processes, and types of warehouses (see section 2.2). Using that knowledge, it is possible to highlight the major differences between the production and warehouse environments (see section 2.3). Finally, it is possible to derive the definition of lean warehousing from the perceptions.

2.1 Genesis of Lean

Discipline and avoidance of waste is deep-rooted in the Japanese culture (Lebra and Lebra, 1986, p. 70). This is even reflected in the daily life of the Japanese. For example, when a Japanese chef filets a salmon to make sashimi he uses the cropped and unused parts of the salmon as ingredients for a soup and does not dispose of them.

Against this cultural background and the aftermath of World War II, Taiichi Ohno assessed the mass production system of the American automobile industry. At that time, the majority of the companies in the automotive industry applied the mass production philosophy strengthened by the circumstance of increasing demands (Ohno, 1988, p.1)(Liker, 2004, p.24). The industry believed that mass production was the most efficient way to fulfill customer demand. Ohno, however, did not share the view of the overwhelming majority.

Ohno saw several forms of waste in the American way and had a strong desire to avoid waste and improve the processes. To serve customer needs with high product variation was one of his desires but this was simply less profitable when mass production was in use. Out of that desire and need he developed the Toyota Production System (TPS) to enable his organization to meet customer demand in a more efficient way. With the Toyota Production System, he established a new culture within Toyota and proved its strength in the oil crisis in 1973. During the oil crisis, which caused decreasing demands, other Japanese industry sectors followed Toyota's method of production (Ohno, 1988; Liker, 2004, p. xiii).

The Toyota Production System received worldwide attention after the publication in 1990 of the Womack book *The Machine that Changed the World* (Womack, 2007). Other publications like Bösenberg and Metzen (1993), Liker (2004), Pfeiffer and Weiß (1994), Rother and Shook (2008), Rother and Kinkel (2009), and Womack and Jones (2004) discussed the lean philosophy from different angles. Dehdari et al. (2011) analyzed several key literature sources and identified the constituents of lean production. The elimination of waste using a structured continuous improvement cycle is the key element in the literature. Furmans and Wlcek (2012) divided this continuous improvement cycle into low and high frequent improvement cycles. The low frequent improvement cycle is an analytic and systematic method resulting from the derivation of the target value streams from their implementation. After implementation, the high frequent improvement cycle stabilizes the implemented standard and improves it again in a systematic and analytical way. In addition to this, Furmans and Wlcek (2012) identified seven success factors for using lean techniques in the warehouse environment. These are leadership, value stream planning, standardisation, work place design, visualisation, work force management, and sustainable problem solving.

2.2 Warehousing at a Glance

Section 2.1 identified the roots of the lean philosophy in the Japanese automotive industry sector. To measure the impact of lean techniques in the warehouse environment, it is necessary to base the research on a definition of the function of a warehouse along with an understanding of the processes within warehouses and the identification of different types of relevant warehouses. This section takes valid definitions for the function of a warehouse, warehouse processes, and types of warehouses from the key literature for the purpose of this thesis.

Gudehus and Kotzab (2012, p. 19) state that the function of a warehouse in a logistics network is to transfer, store and commission goods. Bartholdi and Hackman (2011, p. 5), ten Hompel et al. (2007, p. 50), Arnold et al. (2008, p. 373) use similar definitions for the function. Bartholdi and Hackman (2011, p. 5) mentioned the space and time synchronization function of a warehouse within a supply chain. ten Hompel et al. (2007, p. 50) add the change of status of goods to their definition. Arnold et al. (2008, p. 373) view the warehouse from a broader angle. They assert that the function of a warehouse is, in fact, is to disrupt the supply chain. The European Norm EN 14943 (2005) defines the function of a warehouse as a space that is designed to receive, store, and distribute goods. This thesis will focus on the processes within a warehouse and not on the role of the warehouse within a supply chain. For this reason, this thesis will use the European Norm EN 14943 as the definition for the function of a warehouse.

The three processes mentioned in the European Norm EN 14943 standard are detailed by Arnold et al. (2008, p. 379). In addition to many other information processes, Arnold et al. (2008, p. 379) defined the receipt, storage and retrieval, picking, packing, and shipping of goods as the major processes in a warehouse. Bartholdi and Hackman (2011, p. 24), ten Hompel et al. (2007, p. 53), and Gudehus and Kotzab (2012, p. 19) supported this definition with their own. The VDI 3629 (2005)guideline differs from this definition because it excludes the packing process as a major warehouse process.

Regarding the information processes mentioned by Arnold et al. (2008, p. 379), the VDI guideline overlaps mainly with Arnold et al. (2008, p. 379). Bartholdi and Hackman (2011, p. 28) overlap with the control process and ten Hompel et al. (2007, p. 53) overlap with the identification process. Gudehus and Kotzab (2012, p. 19) do not identify the information processes as major warehouses processes. This thesis will focus on the definition given by Arnold et al. (2008, p. 379) regarding the processes within warehouses because it is the one that is the most validated by the above-mentioned authors and guidelines.

The definition of the different kinds of warehouse types is based on a combination and emphasis of the different major processes. Arnold et al. (2008, p. 376) distinguishes eight categories with 29 warehouse types. The most common warehouse types in the industry sector are production warehouses and warehouses for product distribution. Since the intent of this thesis is to test the hypotheses that are valid for all warehouses, no distinction will be made between the different warehouse types.

2.3 Warehouse versus Production Environment

An innovation is often designed for one specific environment. A wristwatch, for example, was originally designed for use on land. If it is to be used under water, the designer needs to understand the environmental changes and determine if design changes are necessary. The wristwatch, then, must be waterproof and resistant against salt water. It may also possess other features like an altitude meter.

Environmental changes between a production and warehouse environment (see section 2.2) also force us to identify necessary adaptations or design changes to the production-based lean approach (see section 2.1). The environmental changes have to be identified before these adaptations or design changes can be made. Dehdari and Schwab (2012) identified the major differences as follows:

- Differences in the purpose
- Differences in the complexity of problem solving
- Differences in the complexity of movement
- Differences in the order lot size
- Differences in the physical order (space versus line)
- Differences in the expectations of the output performance
- Differences in the leadership

These differences are discussed below.

The purpose in production is to add value to raw materials and change the form; for example, forging steel to make a horseshoe. In warehouses, the added value is to transform the time and spatial status of the product. The trigger for transformation in warehouses is usually an order from a downstream process. This kind of transforming process is called Make to Order (Olhager, 2012). In addition to Make to Order, other possible production triggers are Make to Stock and Assemble to Order.

These differences in purpose mean that problem-solving is usually less complex in warehouses than in production. Changing the form of material is very complex technically and includes several other scientific disciplines. Often only expert knowledge can solve production problems. One example is the difference in complexity in understanding thermal problems in treating materials versus materials handling problems in warehouses, such as closing a box. Difficulties do arise in materials handling problems with getting an overview of the interdependencies between the different processes but production also faces this problem at times. However, it is important to remember that a lower complexity does not mean a simple complexity.

The lot size that is processed in the warehouse is usually one. The reason for this is that when an order is placed by a customer another customer does not usually place exactly the same order. The same order means the same products in the same quantity. The lot size in production can be one but the lot size is often higher to save changeover time or because the production is not mature enough to perform fast changeovers. Higher productivity within the lot can be achieved because of the higher lot size. The higher lot size in production also implies a higher degree of repetitive work for employees and, conversely, a lower degree of repetitive work in warehouses. The lower degree of repetitive work causes higher fluctuation in the working cycles.

Production machines manufacture goods in the same quantity and same quality over a long period of time. This is done because of the lot size and the purpose of adding value to raw materials. Machines are very rare in the warehouse environment. Employees are human beings and, like all human beings, they do not work as precisely as a machine for a long period of time. The result is a higher fluctuation in quantity and quality compared to the work of a machine within the warehouse.

The physical environment of production employees is often structured in the form of a line. One workplace follows another workplace. The workplace is designed with less moving complexity for the worker to ensure higher productivity. The worker usually has a fixed workplace that does not require much walking. In warehouses, these processes are structured in an area that is based on the space required for storing products. This means that the picker has to make different kinds of movements when picking goods from the area. This fact and the lower degree of repetitiveness mean that warehouse employees have to make more complex movements than production employees. In other words, the warehouse environment has longer and higher fluctuating working cycles.

The longer and more volatile working cycles also mean that there are usually no set expectations about the output performance in warehouses. If the estimated level is not reached, the warehouse leader accepts this situation. In production, failure to reach the estimated level will at least result in questions being raised by the leader on the shop floor.

The leadership also changes in the warehouse environment. In production, the leader has an overview of his staff when they are in the production line. Direct communication with all of his workers is possible with only a few restrictions. In the warehouse, the workers are spread all around. Direct communication with the workers is much more restricted. Another problem occurs in the warehouse because the processes are not usually synchronized and the workers change their work and their locations throughout the day. A worker might pick goods in the morning and pack them in the afternoon. This means that the worker reports to two different leaders within one shift. In the production environment, the worker stays at one production station the whole day.

These environmental changes have to be taken into consideration when implementing the lean philosophy in the warehouse environment.

2.4 Lean Warehousing: Transferring Lean Production into the Warehouse

Section 2.1 identified that the key element of the lean philosophy is the elimination of waste using a structured continuous improvement cycle. The environmental changes (see section 2.2) in the warehouse environment (see section 2.3) make it necessary to modify the lean philosophy. More volatile and longer working cycles require an increase in the focus on measuring and controlling the processes. Process controlling has to be done in a systematic and analytical way. The shop floor leaders face the challenge of workers being spread around the warehouse. Since it is not possible to lead the workers directly, a structured continuous improvement cycle has to be considered.

Leadership, measuring, and the driving of improvements in a systematic and analytic way all play an important role when transferring lean approaches from the production environment into the warehouse environment. Dehdari et al. (2011) considered this in his definition of lean warehousing:

> Lean warehousing is a leadership concept. This concept aims at a permanent, systematic, analytic, sustainable, and measurable improvement of processes in the warehouse environment. This happens with the contribution of all employees and with the goal of gaining awareness of perfection in each corporate action.

3 Literature Review: Measuring the Impact of Lean

To measure the impact of lean techniques on performance indicators, it is necessary to compare the development of an indicator for the lean maturity (independent variable) with the development of a performance indicator (assumed dependent variable) for an observed system (see section 1.1). By combining them, it is possible to observe if the lean technique has an impact on performance indicators (see Hypothesis I-III in 1.2). A comparison between two samples is necessary to observe if the lean technique has a higher positive influence on performance indicators compared to other approaches. One sample contains warehouses that focused on the lean approach and the other sample has warehouses without that focus (see hypothesis IV in section 1.2). This measuring concept is equal for production and warehouse environments. This chapter reviews how the existing studies have considered this measuring concept and what kind of measuring tools are available.

3.1 Measuring the Impact of Lean on Production

Womack (2007) performed the most popular effectiveness measurements. In 1990, he compared the performance indicators of Toyota plants with performance indicators of other plants that did not focus on the lean approach. Womack did not measure the maturity level of the Toyota plants. His goal was to show the superiority of lean over other production approaches and he did this by comparing plants that implemented lean with plants that did not.

Table 3.1 shows 14 other research results with a total sample size of 2318. Twelve studies show a positive impact of lean production on the financial performance indicator. Claycomb et al. (1999),Fullerton and McWatters (2001) and Hofer et al. (2012) observed a correlation between a higher lean maturity and the positive development of performance indicators. Biggart (1997) and Jayaram et al. (2008) could not find a statistically significant influence of lean production on performance indicators.

These studies are based on surveys for identifying the maturity of the lean production implementation and to collecting the financial performance indicators (Hofer et al., 2012). One advantage of a survey is that it has a high number of samples that can be analyzed with reasonable resources. A huge disadvantage is that these studies often do not have enough evidence about the reliability of the response data. Often companies do not want to answer surveys, possibly because of the low maturity level of their organizations, or they do not want to take the time to fill out the survey properly. Lean assessments are a more precise way of measuring the maturity of lean techniques (see Doolen and Hacker, 2005) and these are performed by a professional in multi-day workshops.

3.2 Measuring the Impact of Lean on Warehousing

Augustin (2009, p. 94) surveyed the lean maturity of 53 warehouses in his lean warehousing survey. He evaluated the maturity level with just one question using a scale with five maturity levels. Augustin also did not make any statement about the influence of lean

Author(s)	Sample	Dependent variable(s)	Independent variable(s)	Empirical	Findings
Inman and Mehra (1993)	US manufacturing firms adopting JIT (N=114)	ROI, total cost, service	JIT adoption	Regression	Company performance improves as a result of JIT adoption
Biggart (1997)	US manufacturing firms adopting JIT (N=106)	ROA	JIT adoption	Regression	No evidence of a significant effect of lean production adoption on ROA is found
Claycomb et al. (1999)	US manufacturing managers (N=200)	ROS, ROI, profit, profit growth	JIT adoption	Regression	JIT use with customers results in better financial performance
Claycomb et al. (1999)	US manufacturing managers (N=200)	ROS, ROI, profit,	JIT adoption	Regression	The greater the share of JIT transactions, the greater ROI, ROS and company profitability
Callen et al. (2000)	Canadian manufacturing plants (N=100)	Profitability, total costs	JIT adoption	Regression	JIT adoption results in lower costs and higher profits.
Fullerton and McWatters (2001)	US manufacturing firms adopting JIT (N=95)	Profitability improvement	JIT adoption	ANOVA	Greater JIT implementation results in greater profitability improvement
Germain et al. (1996)	US manufacturing managers (N=200)	ROS, ROI, profit	JIT adoption	Regression	JIT results in greater financial performance relative to industry peers.
Kinney and Wempe (2002)	US manufacturing firms adopting JIT (N=201×2)	Profitability, ROA	JIT adoption	Regression	Profitability and return on assets improve after JIT adoption
Fullerton et al. (2003)	Manufacturing firms (N=253)	Profitability, cash flow margin, ROA	Lean production implementation	Regression	Three lean production practice bundles are associated with greater company performance
Matsui (2007)	Japanese manufacturing plants (N=46)	Manufacturing cost	JIT adoption	Canonical correlations	JIT production systems contribute to competitive performance outcomes such as lower manufacturing costs
Jayaram et al. (2008)	Auto parts manufacturers (N=57)	Profitability, ROA	Lean production implementation	SEM	Company performance is not significantly affected by lean production
Fullerton and Wempe (2009)	Manufacturing executives (N=121)	ROS	Lean production implementation	SEM	Lean practices have a direct and mediated positive effect on financial performance
Yang et al. (2011)	IMSS survey data (N=309)	ROS, ROA	Lean manufacturing	SEM	Lean production has a significant positive impact on financial performance
Hofer et al. (2012)	Association for Operations Management in manufacturing organizations (N=229)	ROS, Inventory development, ELI	Lean production implementation	Varimax	Lean production has a positive effect on financial performance

ELI= Empirical Leaness Indicator, JIT=Just in Time, ROA= Return on Assets, ROI= Return on Investment, ROS= Return on Sales, SEM = Structural equationmodeling, US= United States

Table 3.1: Lean Production efficiency studies (Author's illustration based on Doolen and Hacker, 2005) (Inman and Mehra, 1993; Biggart, 1997; Claycomb et al., 1999; Fullerton and McWatters, 2001; Germain et al., 1996; Kinney and Wempe, 2002; Fullerton et al., 2003; Matsui, 2007; Jayaram et al., 2008; Fullerton and Wempe, 2009; Yang et al., 2011; Hofer et al., 2012; ?)

warehousing on performance indicators. Overboom et al. (2010) developed a more detailed assessment for his study that was based on qualitative measures. He published a method for measuring lean maturity that was based on his analyses of Web pages, a questionnaire, and structured interviews of two logistics service providers. However, like Augustin, Overboom did not establish a link between lean maturity and performance indicators. Sobanski (2009) also developed a lean assessment for his warehouses within his study. He verified his assessment and the correlation between subjects with a sample size of 25 warehouses. Standard processes and visual management were two of the areas he studied. He assumed a positive impact of lean on performance indicators for his study. Sobanski (2009) did not test his assumption by relating the assessment results to the performance indicators of the warehouses.

The lack of lean maturity assessments that also consider performance indicators motivated Mahfouz (2011) to develop a new lean assessment for his study. He evaluated a leanness index with a sample size of five warehouses in Ireland. The Mahfouz study was also based on a questionnaire but it included some operational and tactical performance indicators. The cycle time is an example of an operational performance indicator and the number of on-time delivery orders is an example of a tactical performance indicator. Mahfouz used the performance indicators to quantify the results of the lean maturity assessment. He did not analyze the effect of lean approaches on operational or even financial performance indicators.

Augustin (2009), Sobanski (2009) and Mahfouz (2011) concentrated on measuring the level of lean maturity. The level of performance indicator development was considered in the Mahfouz study but only to support the level of evidence of the lean maturity. The Distribution Center Reference Model (DCRM) (see Wisser, 2009) focuses on the level of performance indicators. The DCRM is based on a very sophisticated metric of performance indicators for generating an assertion about the leanness of a warehouse. Unfortunately, the DCRM does not include a metric to evaluate the maturity of the lean techniques.

3.3 Tools for Measuring Lean Warehousing

Data for the abscissa and ordinate is required to test the hypotheses that were defined in section 1.2. Section 3.1 demonstrated that these kinds of comparisons are a mature and standard way of analysing the impact of lean approaches on performance indicators. Section 3.2 showed that there is a gap between the warehouse environment and the production environment in terms of the level of evidence of the lean impact on performance indicators.

Two different measurement tools are necessary to close this gap. The first measurement tool is used to evaluate the maturity of the lean approach in the warehouse environment and the other is used to measure the performance indicators. The existing tools are discussed below.

3.3.1 Lean Warehousing Maturity Assessments

Maturity assessments make it possible to allocate the relative position of a selected domain within a maturity model. The maturity model consists of a set of criteria that are often ordered on a five-point Likert scale. Usually, level one represents the minimum requirement and five represents the highest achievable maturity level (Bruin et al., 2005).

Most assessments verify if a standard has been documented but lack the questions that would verify if the written standard is also executed. If a high level of evidence of the maturity of a selected warehouse is required, it is also necessary to test the execution. Several lean maturity assessments are in use today.

The first step in identifying the most suitable lean maturity assessment for this thesis was to get an overview of the existing maturity assessments. More than 70 maturity assessments were identified by reviewing and researching three scientific databases and the internet and by questioning experts. Seventeen maturity assessments remained after the assessments that did not focus on the lean approach were eliminated. Figure 3.1 shows these 17 assessments. These 17 lean maturity assessments were evaluated again using five criteria with three levels of fulfillment: fully, partially, and not.

The first criterion is Lean Focus. This criterion questioned the depth of lean focus. If, for example, an assessment only asks questions about the Just in Time implementation and no other lean techniques, then this assessment would partially fulfill the first criterion. The second criterion is Verified Execution, which identifies if the maturity assessment verified the execution of the lean approach. This is related to the point mentioned earlier that most assessments only determine if a standard is documented. The third criterion is Not Survey Based and this examines the collection of the data. The data is more reliable and objective if it is not survey based and if it is provided by different individuals in the warehouse. The fourth criterion, Warehouse Focus, determines the depth of the focus on warehouse operations. For example, this criterion is used to determine if the assessment evaluates the warehouses processes (see section 2.2). The fifth criterion is Tested in Practice and the purpose of this criterion is to determine if the assessment asks questions about the testing of lean techniques in practice.

Fullerton et al. (2003) focused on the lean approach but they only covered the Just in Time technique with their research and missed others. Fullerton et al. (2003) did not focus on warehouse operations and instead focused on the production environment.

The Lean Enterprise Self-Assessment Tool (LESAT) was developed by researchers of the Massachusetts Institute of Technology (MIT). The LESAT is based on the Capability Maturity Model for Software (CMM) and focuses on the lean enterprise and not, specifically, on the warehouse environment. The strength of the CMM is that it is not survey based and it was developed by science for use in

	Lean Focus	Verified Execution	Not Survey Based	Warehouse Focus	Tested in Practice
Fullerton, McWatters and Fawson (Fullerton et al., 2003)	◐	○	○	○	●
Lean Enterprise Self-Assessment (Hallam, 2003)	●	○	●	◐	●
Perez and Sanchez (Perez et al., 2000)	●	○	●	○	◐
Panizzolo (Panizzolo, 1998)	●	○	●	○	◐
Shah (Shah, 2003)	●	○	○	○	●
Jordan and Michel (Jordan et al. 2001)	●	○	○	○	◐
The 360° Lean Audit (Dollen et al. 2003)	●	○	●	○	◐
Lean Company Survey (Dollen et al. 2003)	●	◐	○	○	◐
HPEC Assessment (Dollen et al. 2003)	●	◐	●	○	◐
Lean Checklist Self-Assessment (Dollen et al. 2003)	●	○	◐	○	◐
Lean Business Assessment (Dollen et al. 2003)	●	○	◐	○	◐
How Lean is Your Culture? (Dollen et al. 2003)	◐	○	◐	○	◐
Dell Business Assessment (Shan, 2008)	●	○	●	○	●
CMMI for Services (CMMI Product Team, 2010)	●	○	●	○	●
Overboom (Overboom , 2010)	●	○	●	◐	●
Sobanski (Sobanski, 2011)	●	○	●	●	●
Mahfouz (Mahfouz, 2011)	●	○	●	●	◐
BPS Assessment V. 3.1. (Bosch2012a)	●	●	●	◐	●

● Full fulfilled   ◐ Partial fulfilled   ○ Not fulfilled

Figure 3.1: Lean Maturity Assessment overview (Fullerton et al., 2003)(Hallam, 2003; Pérez and Sánchez, 2000; Panizzolo, 1998; Shah, 2003; Jordan and Michel, 2001; Doolen and Hacker, 2005; Shan, 2008; CMMI Product Team, 2010; Overboom et al., 2010; Sobanski, 2009; Robert Bosch GmbH, 2012)

the aerospace industry. It has also been verified and modified several times (Hallam, 2003). Pérez and Sánchez (2000) and Panizzolo (1998) used field-based surveys but they did not cover all warehouse operations. They verified their theory with the help of a small sample size in Spain and Italy. Shah (2003) and Jordan and Michel (2001) did have higher sample sizes but their research is based on the survey-based approach. They also did not focus on warehousing operations.

In addition to six scientific research-based lean assessments that are also considered in this section, Doolen and Hacker (2005) described several lean maturity tools that were developed and used in the industrial environment. None of them focused on warehouse operations but two of them partially verify the execution of the lean approach. The Lean Company Survey and HPEC questioned the role of the performance indicator to determine the outcome of lean implementation. Shan (2008) and the CMMI Product Team (2010) also did not focus on warehouse operations but these assessments are often used in practice.

As discussed in chapter 3, Overboom et al. (2010), Sobanski (2009), and Mahfouz (2011) developed lean warehousing assessments. Their assessments failed to cover major warehouse operations (see section 2.2) or were not conducted with a large enough sample size (Mahfouz, 2011). In addition to this, their research does not determine if a standard has been executed. This is missing in all of the assessments that have been analysed so far.

Robert Bosch GmbH (2012) developed the only lean maturity assessment that has a focus on the existence and execution of implemented lean techniques and verified them with performance indicators. This assessment has been used multiple times in more than 290 plants in different business sectors all over the world. This assessment is not survey based. Unfortunately, it focuses on production and only covers some of the warehouse processes that are mentioned in section 2.2.

3.3.2 Lean Warehousing Performance Indicators

To measure the impact of lean approaches on performance indicators, we need to measure performance indicators in addition to measuring the lean maturity. Depending on the intensity of the efforts for implementing lean techniques into a warehouse, the impact could vary when measuring a performance indicator that includes all areas within a warehouse. In the beginning of a lean journey, some warehouses might implement the lean approach in isolated warehouse areas. In this scenario, a measurement that uses a performance indicator that covers all warehouse areas would not represent the true effect: the effects of other areas without lean implementations would influence or even overlap the effect from the selected area.

For example, the decision is made to standardize the processes in a warehouse and a shop floor cycle with workforce management is implemented in a picking area. Measuring the performance of the entire warehouse and drawing conclusions about the lean impact would not represent the true effect. This is because the effects of other areas influence the overall performance indicator. It is rather like seeking to measure the heat of a small flame located on one side of a room but doing so by measuring the room temperature on the other side and concluding that the flame does not affect its environment even though the temperature close to the flame is high.

A system is required that will measure precisely at a specific level and cover the effect that different performance indicators have on each other. Key Performance Indicator (KPI) trees fulfil this requirement (see Scheer, 2005). These kinds of KPI trees are in use at Bosch and are named in the Bosch methodology as Bosch Key Performance Result (KPR) and KPI Trees. The structure of the Bosch KPR/KPI Tree is described in figure 3.2. In general, it has four key performance levels: the top KPR, the value stream KPR, the monitoring KPI, and the improvement KPI level.

The result KPR level includes the top KPR for a selected warehouse, involving such aspects as total warehouse costs. The value stream

KPR level describes the costs for the different value streams within the warehouse, including such aspects as the cost for distributing a full pallet. The KPI for a specific area within the warehouse is measured at the monitoring level. For example, the KPI could evaluate the productivity of the packaging area or the value stream for the cost for distributing a full pallet. The most detailed measuring is done at the improvement level. These measurements could include the quantity of pallets that packer one is packing today. The KPI at the improvement level are usually used for concrete improvement work.

All of these levels are linked with each other. Changes in one level will be transmitted to the other levels. For example, the productivity KPI of one packaging team will influence the productivity KPI of the total packaging area and even the total cost KPR of the entire warehouse but with less intensity because other teams, like the pickers, also influence the total cost of the warehouse.

The purpose of this thesis is to measure the impact of lean approaches on performance indicators. By measuring this, we have to consider that lean is not something that decision makers want to have and they can buy and then it is done. Usually, a seed has to be planted in a specific area of the warehouse. If the people around this seed take care of it, it will grow and spread around the whole warehouse. This will take time and, as long it is not spread around the whole warehouse, the specific area where it grows has to be measured. The effect of this area might have such a big impact on the KPR that it can be noticed by observing the KPR.

A KPR/KPI Tree can be used to recognize the impact in the different areas and at the different levels. The four levels of the KPR/KPI Tree make it possible to measure the impact of implemented lean techniques with different levels of penetration.

The full implementation of a KPR/KPI Tree with all four levels within a warehouse is rare today. The KPIs for the improvement level are usually used for concrete improvement work and are often not measured constantly. The value stream KPRs are difficult to

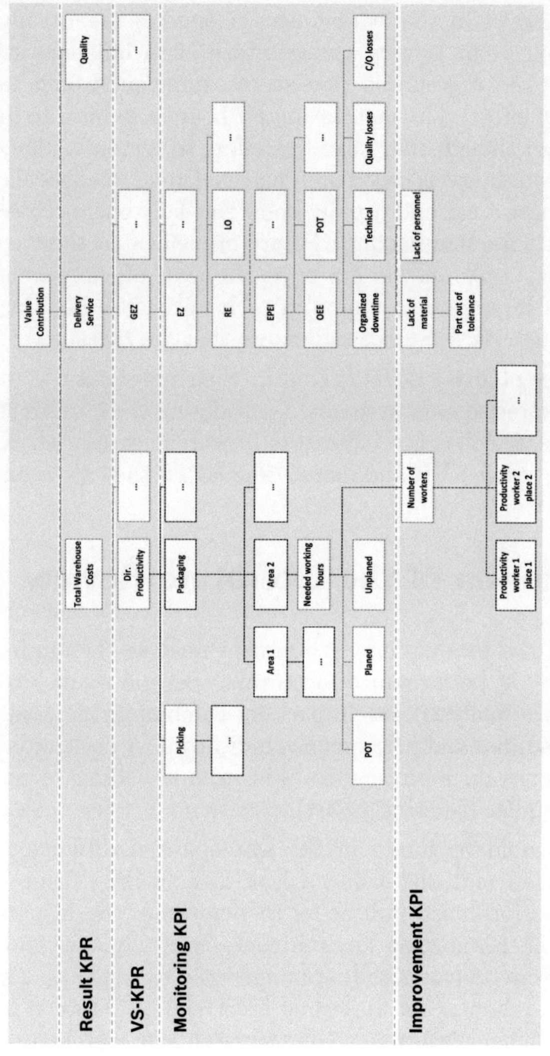

Figure 3.2: KPR/KPI Tree example

measure within warehouses because several, if not hundreds, of value streams are merged in the warehouse. Resources are usually not dedicated to one value stream and one task often influence several value streams. As a result, a precise measurement would take a huge amount of effort. The monitoring KPIs are also hard to find in warehouses even though they take less effort to measure. Since the structure of a warehouse is based on functional areas, leaders allocate capacities to these areas and usually know the daily output. Because this is known, if the monitoring KPI are implemented they usually still focus on one area and do not cover all areas of the warehouse. Even result KPRs are not common in each warehouse but they are the most common performance indicators that are measured.

A fully implemented KPR/KPI Tree in each warehouse would be desirable for a precise measurement. Considering the current status of the available performance indicators in warehouses, it is realistic to focus on the result KPR and monitoring KPI if they are available.

3.4 Conclusion of the Literature Review

The studies in the production environment analysed the impact of lean techniques on performance indicators. Several studies with a high sample size analysed the impact by combining the results of lean maturity studies and performance indicators. A positive impact of lean techniques on performance indicators is backed by several independent studies (see section 3.1).

Some studies could be found in the warehouse environment that analysed the lean maturity using a lean assessment. Other studies analysed performance indicators to determine the lean maturity. Unlike the production environment, no study was found in the warehouse environment that combined the two factors: neither in pure lean warehouses for analysing a correlation between higher maturity and higher performance nor between lean warehouses and warehouses without lean techniques for analysing how each group performs (see section 3.2).

In addition to this, the existing lean maturity assessments for warehouses do not fulfil the desired minimum criteria (see subsection 3.3.1). The Bosch Production System Assessment V. 3.1 has the highest level of fulfilment of the criteria but it does not focus solely on warehouse operations.

In conclusion, there is a huge gap between the levels of evidence for the impact of lean techniques on performance indicators in the different environments. It is not currently possible to determine the impact of lean approaches on performance indicators within the warehouse environment. Thus, a new study is needed to close this gap in evidence and analyse the hypotheses described in section 1.2.

Unfortunately, none of the existing lean maturity assessments that focus on warehouses are adequate enough to fulfil the desired minimum criteria. The Bosch Production System Assessment V. 3.1 has the highest level of fulfilment of the criteria but it focuses on production and only covers some of the warehouse processes as mention earlier. An adaption of this assessment for the warehouse environment would fulfil the desired criteria and enable further studies within this subject.

4 Bosch Logistics Warehouse Assessment

Within the scope of this study, the Bosch Production System Assessment V. 3.1 was adapted for the warehouse environment and transformed into the Bosch Logistics Warehouse Assessment (BLWA). The BLWA was developed for the Warehouse Excellence project of Bosch in the beginning of the year 2010 (compare also Dehdari et al. 2012). Several sources and experts were consulted for the transformation of the Bosch Production System Assessment V. 3.1 into the BLWA.

The company Bosch and the Warehouse Excellence project will be discussed later in chapter 5. This chapter describes the development and then the structure of the assessment. The purpose of this chapter is to give a general overview of how the assessment works.

4.1 Development of the Bosch Logistics Warehouse Assessment

The BLWA was developed in several steps. In the first step, the key literature for the lean approach was re-evaluated (see section 2.1). The main components of the lean approach that needed to be assessed in the warehouse assessment were identified based on the defined requirements. The existing lean maturity assessments that focused on the warehouse environments (see section 3.2) were also analyzed. The goal was to find the components that could be used for the Bosch Logistics Warehouse Assessment.

The second step involved the creation of a new structure. The Bosch Production System Assessment V. 3.1 only focused on some warehouse operations. The warehouse processes (see section 2.2) that were not covered by the processes in the Bosch Production System Assessment V. 3.1 were included. Matched with the content of the literature and other assessments, the first draft of the BLWA was finalized.

The first draft of the BLWA was reviewed by experts from the Bosch Production System. After their feedback was included, guideline-based interviews were used to have the first draft checked by experts from several organizational levels within Bosch. These experts included representatives from the corporate level, the business unit, warehouse leaders, and shop floor personnel. Experts from the Karlsruhe Institute of Technology were also questioned.

A test version of the BLWA was released after the feedback from the interviews was incorporated. The test version was tested in three warehouses. The feedback from the test version was taken into consideration and, after a final review by the corporate Bosch Production System expert team, the BLWA was released.

4.2 Structure of the Bosch Logistic Warehouse Assessment

The structure of the BLWA is shown in figure 4.1.

The structure is divided into three segments: the Continuous Improvement Process (CIP), overall subjects, and warehouse processes. The CIP consists of the System-CIP and Point-CIP, which are Bosch-specific terms that were developed by Bosch Production System experts (Robert Bosch GmbH, 2012).

The System-CIP pertains to process and value stream design. It aims to capture the current value stream status with techniques

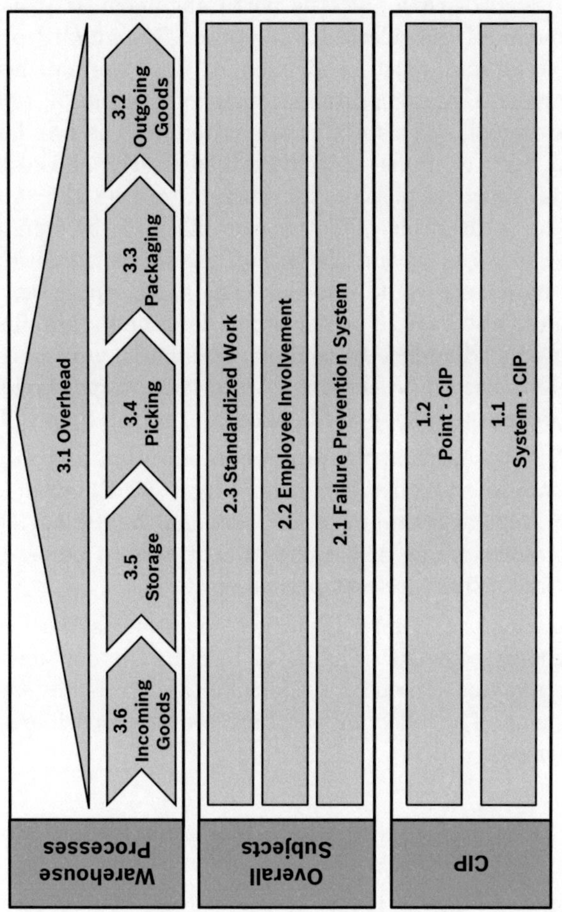

Figure 4.1: Bosch Logistics Warehouse Assessment structure

like value stream mapping (Rother and Shook, 2008), layouts, and spaghetti diagrams (Flinchbaugh, 2009). The value stream target can also be designed using the true north alignment and the business requirements of the selected warehouse. System-CIP projects with target conditions must be defined to close the gap between status and target. Target conditions consist of a standard, a performance indicator goal, and a stabilization criterion. The standard has to be defined, easy to understand, described clearly, and displayed on-site. It also must be possible for workers to meet the standard and it must be measurable. The measurability of the standard is important because the second element of the target condition, the performance indicator, would not make any sense otherwise. Comprehensive limits also have to be defined to describe the stabilization criterion. Finally, adherence to common standards, guidelines, and legal restrictions has to be ensured. Once these target conditions have been implemented, they are handed over to the Point-CIP.

The Point-CIP is a method for process stabilization and improvement. It is comprised of five elements: target condition, quick reaction system, regular communication, sustainable problem solving, and process confirmation. This method continues to be used until the stabilization criteria are met permanently.

The definition for the Point-CIP target condition is the same as the one for the System-CIP. The quick reaction system consists of a trigger for reactions, defined responsibilities, measures, and an escalation scheme. The following questions are used to define a quick reaction system:

- Who? - Responsible person for taking action – starting at the operator level
- How long? - Time limit for problem solving at each level
- What? - Systematic description of measures, how to document facts
- How? - Problem solving method to be used

- Output? - Records to be created (for example, problem solving sheet)
- Valid? - 24 hours, Monday to Sunday

Regular communication consists of the definition of the participants, the tools used, and a schedule. Regular communication can be established at many points within a warehouse: it can be defined at different hierarchical levels or in different areas. An aligned sequence of the various types of regular communication supports the input and output of information for the different meetings. Regular communication supports the process of sustainable problem solving and the exchange of information between all departments by rules defined at all levels.

The following elements are important for sustainable problem solving: a root cause analysis, sustainable countermeasures, sustainable proof of rollout to other areas, prevention of re-occurrences, and a standardization of the result. Sustainable problem solving should be done in a team with problem solving experts, leaders from the area, and shop floor workers.

Process confirmation is a verification of the adherence of the operators to a standard. It also contains an analysis of any deviations from the standard that occurred. Standards have to be checked frequently because of fluctuating process outputs, varying parts (for example, changing the combination of parcels), and changing operators (compare also section 2.3).

The strength of the System-CIP and the Point-CIP is the linking and combination of the elements with each other. The following real-world example will make this clear. The System-CIP cycle highlighted that a new milk run was necessary in one of Bosch's production warehouses in the south of Europe. The target condition for the new milk run was defined: a timetable, a cycle time of 16 minutes for the route, and plus or minus 1 minute as the stabilization criterion. After the milk run was implemented and the workers were trained, the target condition was handed over to the Point-CIP. Hours after the implementation, the milk run driver reached a station on his route with a 1.5 minutes delay. The quick reaction system helped the milk run driver react accurately. He escalated the delay to his supervisor and asked for the support of another worker. The quick reaction system also defined that he had to document the delay using the questions mentioned above in a regular communication. In the next regular communication, the supervisor asked the driver for the information about the delay of the milk run and because this was a newly defined critical implementation he decided to establish a problem solving team to understand why the milk run came late. After investigating the problem in a structured way, the team discovered that the milk run collided with another milk run every second or third cycle. They performed some tests and proposed a solution to reschedule the new milk run. After discussing this proposal in the standard management meeting, the standard of the milk run was changed. As part of the process confirmation, the cycle time adherence was checked daily for three months. The problem was only considered to be solved permanently if the milk run did not come late again during this time period.

The roots of the System-CIP and the Point-CIP are in the Japanese automotive production systems. A lot of the lean systematic mentioned in section 2.1 by various authors, and in particular, analyzed and summarized by Dehdari and Schwab (2012) is covered by System- and Point-CIP. For example, the general topics of failure prevention systems, employee involvement, and standardized work. The main warehouses processes in the assessment represent the processes described in chapter 2.2. Only input and output are included under storage. A detailed list of the topics that are covered is shown in the figures 4.2 and 4.3.

The topics also have subtopics. To cover each subtopic, several criteria have to be assessed with different maturity levels. The maturity levels start at 0 and go up to the level 4. The higher the level, the more challenging and mature the criteria. Each criterion has a con-

| | 1. CIP | | 2. Overall Subjects | | |
	1.1 System-CIP	1.2 Point-CIP	2.1 Failure Prevention System	2.2 Employee Involvement	2.3 Standardized Work
Conception	Business requirements	Target condition	Work Content	Involvement	Coverage of Standardized Work
Conception	Value Stream planning	Quick reaction system	Visualization	Target Deployment Team Lead	Visualization
Conception	Identification of improvement activities	Regular communication		Qualification / Training	Qualification
Conception	Definition of target conditions	Sustainable problem solving			
Conception	System-CIP projects	Process confirmation			
Conception	Point CIP areas				
Execution	Target derivation	KPI-effect	Error rate	Multi-skilled Operators	Stability
Execution	System CIP cycles	Quality of problem solving		Operator Involvement	5S Status
Execution	Improvement focus			Leadership Involvement	Productivity
Execution	Leadership involvement				
Execution	VSM-Quality				
Execution	Target achievement				

Figure 4.2: Bosch Logistic Warehouse Assessment Topics Part 1

	3. Warehouse Processes					
	3.1 **Overhead**	**3.2** **Outgoing Goods**	**3.3** **Packaging**	**3.4** **Picking**	**3.5** **Storage**	**3.6** **Incoming Goods**
Conception	Qualification	Organization	Packaged Good	Picking Process	Storage Technic/Layout	Receiving Process
		Technical Equipment	Packaging Material	Organizational System	Storage Criteria	Entry/Booking
		Visualization	Packaging Process	Information System	Inventory Management	Inspection
			Visualization	Visualization	Visualization	Visualization
Execution		Time window adherence	Lead time of the packaging process	Lead time of the picking process	Lead time of the storage process	Time window adherence of receiving
		Balancing of complete shipping processes	Packaging error rate	Picking error rate	Storage error rate	Balancing of receiving processes
		Lead time of the shipping process				Lead time of the receiving process
		Dispatch error rate				Receiving error rate
		Handling steps				Handling steps

Figure 4.3: Bosch Logistic Warehouse Assessment Topics Part 2

cept dimension and an execution dimension that are linked together. This link ensures that the documented standard will be checked to see if it is executed. Figure 4.4 explains the relationship between topics, components, criteria, and maturity levels. The link between concept and execution is explained by the example in figure 4.5. In this figure, the standard for time windows is asked for at the conceptual stage. During execution, the adherence to the time window is sought.

4.3 Intermediate Result: Measuring Systematic

To evaluate the impact of lean techniques accurately, we need to measure the maturity of the lean entity and the performance change with performance indicators. Chapter 4 identifies a gap between the maturity of the methodology measuring the impact of leanness in a production environment and that in a warehouse environment. Detailed measurements are needed to gain reliable knowledge. This is why section 3.3.1 identifies BPS Assessment V3.1 as the most precise in terms of the requirements. However, an adaption was necessary because it was designed for the production environment. The BLWA fulfills these adaptation needs. Together with the KPR/KPI Tree (see sub section 3.3.2), accurate measurement techniques are now available to test the hypotheses described in section 1.2.

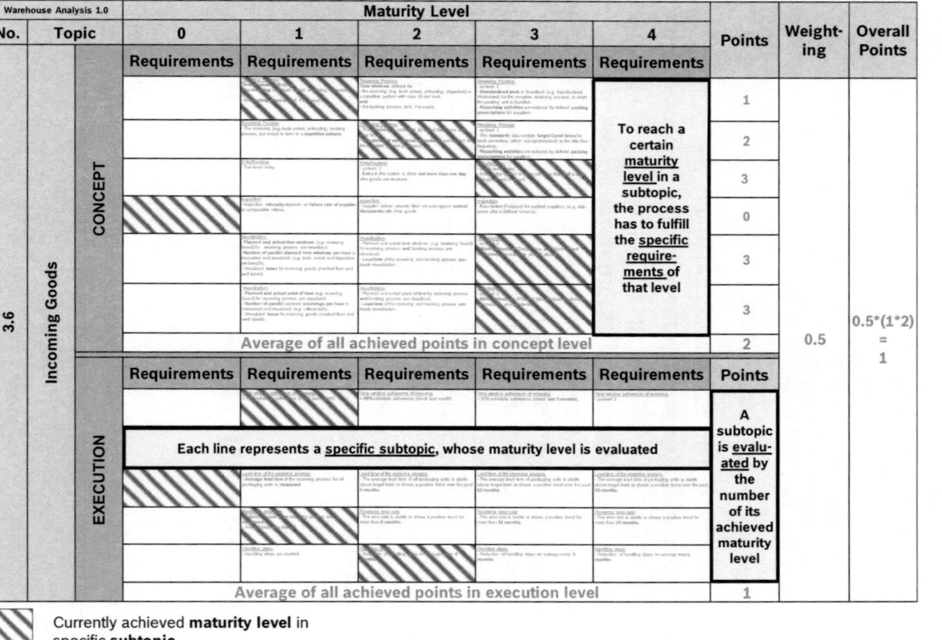

Figure 4.4: BLWA link between execution and concept

Warehouse Analysis 1.0		Maturity Level					Points
No.	Topic	0	1	2	3	4	
		Requirements	Requirements	Requirements	Requirements	Requirements	
3.6	Incoming Goods — CONCEPT		Receiving Process: **Time windows** defined for: • The **receiving** (e.g. truck arrival, unloading, departure) Or: • The **booking** process (including puting into storage)	Receiving Process: **Time windows** defined for: • The receiving (e.g. truck arrival, unloading, departure) in a repetitive pattern with max. 3h per truck **And:** • The booking process (including puting into storage)	Receiving Process: • As level 2 • **Standardized work** is described (e.g. Standardized Worksheet) for the complete receiving process • **Repacking activities** are reduced by defined **packing instructions** for suppliers	Receiving Process: • As level 3 • Operators are working **>90%** of their work time according standardized work	1
		…	…	…	…	…	…

Warehouse Analysis 1.0		Maturity Level					Points
No.	Topic	0	1	2	3	4	
		Requirements	Requirements	Requirements	Requirements	Requirements	
3.6	Incoming Goods — EXECUTION		Time window adherence of receiving: • **> 70%** schedule adherence (check last month)	Time window adherence of receiving: • **> 90%** schedule adherence (check last month)	Time window adherence of receiving: • **> 90%** schedule adherence (check **last 3 months**)	Time window adherence of receiving: • As level 3	1
		…	…	…	…	…	…

Figure 4.5: BLWA example for linked criteria

5 Design of the Experiment

The previous chapters identified and developed the measuring method. This chapter discusses the design of the experiment of this study. This means that we get to know the structure of the observation sample and the control sample that are parts of the design of the experiment. This includes the sample sizes, the warehouse types (see section 2.2) and geographic regions of the warehouse locations. The use of the measuring methods identified and developed in section 4.3 are also described. The analysis of the samples in this chapter should make it possible to determine the validity of the defined hypotheses (see section 1.2) in chapter 6.

5.1 Warehouse Excellence Group - the Observation Sample

The Bosch Group is a supplier of technology and services in the areas of automotive and industrial technology, consumer goods, and building technology. These broad areas form a good cross section of the economy. The Bosch Group is made up of Robert Bosch GmbH along with its roughly 350 subsidiaries and regional companies in some 60 countries, including over 800 warehouses. Nearly half of the Bosch warehouses are operated by Bosch and the other half are run by logistics service providers.

The performance of these warehouses is crucial to the success of the company. High delivery performance targets and quality requirements have to be fulfilled. These warehouses also cause significant costs. However, a Bosch internal observation study showed a gap in the lean maturity level between the production and warehouse environment. In order to close this gap, the Bosch Group established a pilot project to test and evaluate the adaptation of the Bosch Production System to the warehouse environment. The Bosch Group also decided that an additional goal of the project was to measure the impact of the Bosch Production System on performance indicators. This impact measurement would form the basis for the decision for a worldwide rollout that would affect all 800 warehouses.

The Bosch Group named the pilot project Warehouse Excellence. The Warehouse Excellence group was chosen randomly and consisted of 16 warehouses located in seven countries. These were comprised of six distribution warehouses, seven plant warehouses, and three raw material warehouses. These 16 warehouses handle three different kind of businesses. The different businesses are automotive technology goods, industrial technology and consumer goods, and building technology. Fourteen warehouses out of the 16 warehouses are single user warehouses and handle one business. Two of the 16 warehouses are multi-user warehouses and each of them handle two different businesses. In total, two are involved in industrial technology, 10 warehouses handle automotive technology, and six deal with consumer goods and building technology.

Eight of the warehouses were operated by Bosch and another eight by logistics service providers. These included three of the five biggest logistics service providers in the world as measured by the turnover. All other detailed structural information are given in appendix A and appendix B.

5.1.1 The Warehouse Excellence Project - Lean Empowerment

The project ran from November 2010 to March 2012. Key performance indicators are available from January 2010 to December 2011. It is safe to assume that there was less focus on lean in the year 2010 than in the year 2011. So, ideally, these two years can be compared with each other. This section describes the empowerment program, milestones, and the available data that was gathered during this period.

Empowerment

The actions undertaken within the warehouses were based on an empowerment program. Figure 5.1 shows the four elements of the empowerment program. The empowerment program is based on the literature in section 2.1 and the experience of Bosch Production System experts. The aim is to enable the warehouse leader to drive the continuous improvement process as per the lean warehousing definition given in section 2.4. As described in chapter 4, the System-CIP and Point-CIP play a significant role in the achievement of that goal. The empowerment program consists of four elements that are described below.

Each of the three Bosch Logistics Workshops (BLW) was held over two days. The workshop sought to introduce knowledge about the System-CIP (BLW I), the Point-CIP (BLW II), and special problem solving (BLW III) in a practical way. For example, the theory behind value stream mapping was taught in an hour-long classroom lecture. The participants then tried value stream mapping within instuctor-led groups in the warehouse. A separate session was necessary for problem solving because it is so highly complex. The first workshop took place in December 2010, the second in March 2011, and the last in June 2011.

The Bosch Logistics Learning Groups (BLLG) helped warehouse managers solve upcoming problems together. Rotational one and a half day visits to different warehouses also supported knowledge transfer between warehouses. The participants met six to seven times during the project period.

The Bosch Interdisciplinary Local Teams (BILT) supported knowledge transfer from the workshops to the warehouse operations of the participants. Lean experts coached warehouse managers to adjust and implement the learned methodologies. The coaches also provided detailed feedback to those in leadership roles.

The final component was the exchange of good practices with the lean management working group in the warehouses of the German Logistics Association (BVL). In this working group, several companies identified the lean success factors and tested this out in pilot warehouses of their own. Furmans and Wlcek (2012) summarized and published the results.

Milestones

The milestones provided a direction and challenged the warehouses within their lean activities. These milestones were to be reached during the project period. The first milestone required value stream mapping, value stream design, and a project plan. The project plan was to close the gap between the two value streams. These represented parts of the System-CIP cycle. The second milestone involved establishing regular communication and visualizations. The latter included the visualization of standards, process statuses on the shop floor, and major key performance indicators. The second milestone served as preparation for the third, which required one closed Point-CIP cycle from the participating warehouses. In addition to regular communication and target conditions, this included three other elements: a quick reaction system, problem solving methods, and process confirmation. Figure 5.2 shows the milestones covered by the empowerment components.

Measuring

Three different measurement techniques were used to produce a holistic picture of warehouse performance. These included the BLWA, result KPR tracking, and monitoring KPI tracking if available.

In order to reach the milestones, the lean maturity of the warehouses

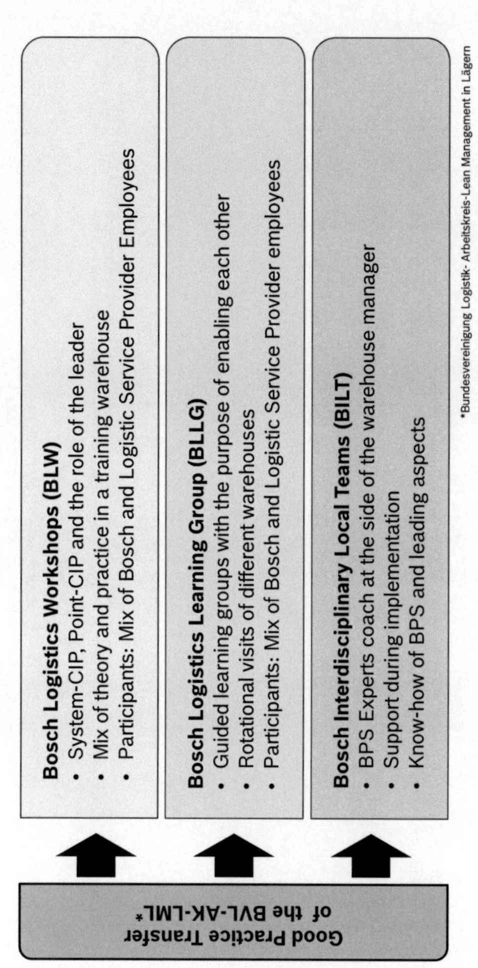

Figure 5.1: Warehouse Excellence empowerment structure

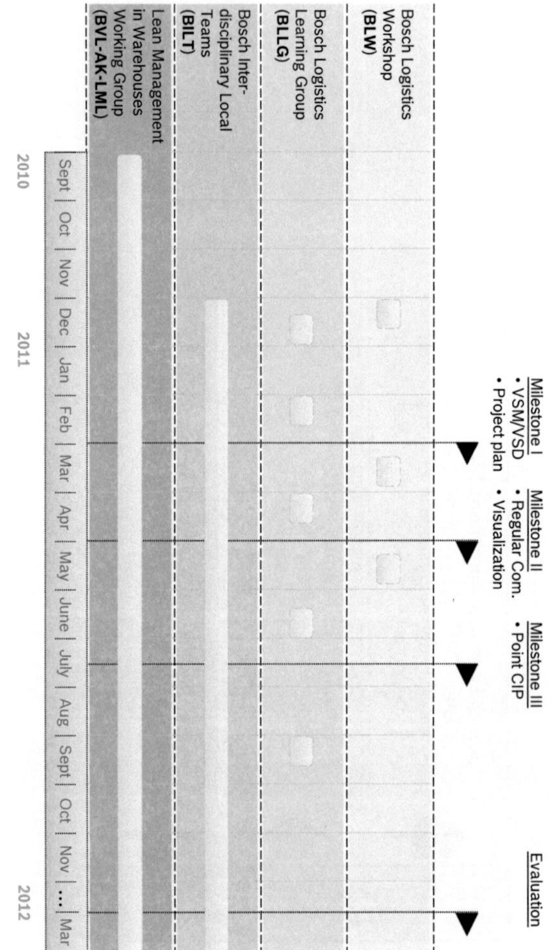

Figure 5.2: Warehouse Excellence project milestones

was recorded with the BLWA (see chapter 4) before warehouse activities began. The maturity in the 70 components (see figure 4.2 and 4.3) were recorded by the assessors for each warehouse. By multiplying the number of components with the number of warehouses, we find 1120 ordinal data at our disposal (compare Bortz and Weber, 2005, for a definition of ordinal data). Each record was made on the shop floor by two assessors. Feedback rounds were used to ensure that the observations were aligned between the assessors and the warehouse leaders. This ensured that subjective perceptions were kept to a minimum. The assessors used a laboratory book as an organizational tool to record the most important environmental points not covered in the structure sheet.

Rational data (Bortz and Weber, 2005) were provided by the monitoring KPI measurements. Result KPRs from January 2010 to December 2011 were also analyzed. Delivery performance, quality performance, and productivity were also taken into consideration.

Delivery performance at the result KPR level represents the overall process in most organizations. Delivery performance usually only measures if the confirmed customer order was fulfilled correctly at the end of the supply chain. To know the root cause, it is necessary to know where the problem occurred within the supply chain. For example, one frequent reason for a delivery performance error is that the product was not assembled in time. A more detailed measurement, such as that between two internal processes, would indicate where it happened but this kind of measurement is rare. This indicates that customer claims frequently cannot be tracked back to the supply chain participant that caused the problem.

Another point regarding the measurement of delivery performance is that very often no trend or fluctuation could be recognized in the measured performance indicator within the project. The reason for this is that the workforce in the consumer goods warehouses of the Warehouse Excellence group is highly flexible. The workforce usually extends their working hours until the last order is fulfilled. A flat delivery performance was also common in the participating automotive warehouses. No recorded incident suggested that a customer delivery performance claim was caused by the warehouse. It will be difficult to conclude any impact of the lean approach on this performance indicator at the result KPR level because of that overall measurement of the delivery performance indicator and the usually flat line of the delivery performance. To measure the impact within the Warehouse Excellence project, a more detailed measurement is required than what is available now.

Like with delivery performance, the quality performance indicator at the result KPR level also measures the overall process. The measurement shows if the right product in the right quality and in the right quantity reached the customer of the end of the supply chain, regardless of where the mistake was made. There is also a time gap between the error caused on the shop floor and the reported claim from the customer. This time gap often blurs the KPR. In the automotive sector, the annual claim rates are at most a single-digit figure. This statistic of rare events also frequently leads to a flat quality performance line. To measure the impact within the Warehouse Excellence project, a more detailed measurement of the quality performance indicator is required than what is available now.

Productivity is measured by comparing an output factor with an input factor. In the warehouse, a common measurement is to divide order lines by man-hours. This means that the measured performance indicator can represent the focused area exactly. Gaps in time do not exist between the effect caused on the shop floor and the effect discovered with the KPI. Fluctuating order volumes and adjustments in the workforce also lead to volatile productivity lines that support effect analysis. Measurements between two different internal processes using monitoring KPI are also possible and partly available in warehouses. The lack of time gaps, volatile productivity lines, and the availability of monitoring KPIs are all sufficient for measuring this performance indicator for this thesis.

Another reason why no trend can be recognized by observing the quality performance and delivery performance indicators are the boundary conditions. For a warehouse operation, quality and delivery performance is more a constrain and productivity is a target. This means that the priority is clear and to the account of productivity performance that let a trend be recognized.

In summary, detailed measurements of the delivery performance and quality performance figures will not be taken into consideration. Of course it is important to know that the delivery performance and quality performance have not been influenced negatively by the lean approach so far. We know that no major incident has taken place within the project period. A detailed measurement like the productivity performance measurement is not possible at this moment because of the above mentioned reasons. Therefor the focus will be on productivity measurements. Productivity figures from January 2010 to December 2011 are available for each warehouse. Monitoring KPI from areas that focused on implementing lean are also available and will be analyzed. The data for analysis usually ranges from a period before implementation to a certain period after it. The minimum length of time considered before the lean implementation for control measurements is usually, at a minimum, the same length of time considered for measurements after implementation.

5.2 Control Group

A control group was needed to test hypothesis IV in section 1.2. The control group consisted of 56 randomly selected Bosch warehouses located across 16 countries. These included 38 distribution warehouses and 18 plant warehouses. Thirty-seven warehouses are in the automotive business, 6 in industrial technology, and 13 in consumer goods and building technology. Twenty-seven of the warehouses are operated by Bosch and 29 by logistics service providers. This included three of the five largest logistics service providers in the world as measured by the turnover. All other detailed structural information is provided in appendix B.1. The control group was not influenced by an empowerment program or by milestones that they had to achieve.

Measurements were carried out that were similar to those undertaken for the Warehouse Excellence group. At the same time, leanness measurements in the control group were conducted using an online survey. Before the online survey was released, it was tested on seven staff members from the Karlsruhe Institute of Technology (KIT), one lean expert, and two warehouse managers. The diversity of the test participants ensured feedback from different perspectives. The questionnaire was finalized after the feedback from the test was taken into consideration.

To ensure high response quality, the participating Bosch warehouses were encouraged by Bosch's logistic steering committee to complete the survey. The project team contacted each warehouse manager to explain the purpose and structure of the survey. The warehouses were given two weeks and the support of the project team to complete the survey. A plausibility check ensured that the survey answers were filled in precisely. For example, if a warehouse did not have value stream mapping, they could not have a project plan that included a value stream status and a value stream target. Such issues were handled by the questionnaire. The final questionnaire consisted of 19 components that covered the System-CIP and the Point-CIP. A total of 1064 ordinal data were at our disposal.

As a result of the discussion in subsection 5.1.1, the control group also focused on productivity KPI so that rational numbers could be acquired. From the 56 randomly chosen warehouses, only 18 warehouses measured productivity at the result KPR level. This is why only 432 rational numbers were acquired.

5.3 Method for Testing the Hypotheses

The method for testing the hypotheses I-IV in section 1.2 is described in this section. The verification structure is presented in figure 5.3.

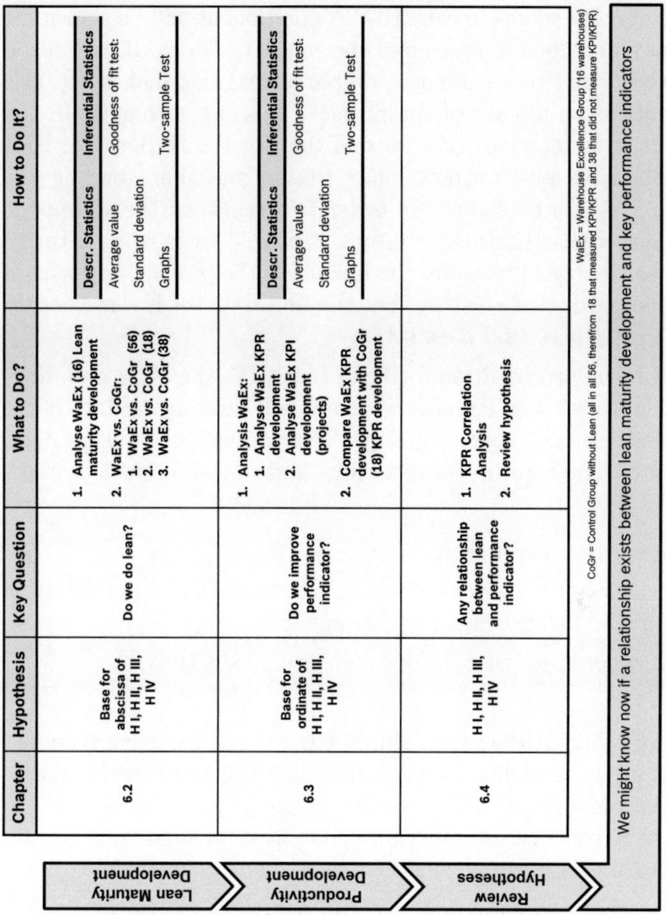

Chapter	Hypothesis	Key Question	What to Do?	How to Do It?	
6.2	Base for abscissa of H I, H II, H III, H IV	Do we do lean?	1. Analyse WaEx (16) Lean maturity development 2. WaEx vs. CoGr: 　1. WaEx vs. CoGr (56) 　2. WaEx vs. CoGr (18) 　3. WaEx vs. CoGr (38)	**Descr. Statistics** Average value Standard deviation Graphs	**Inferential Statistics** Goodness of fit test: Two-sample Test
6.3	Base for ordinate of H I, H II, H III, H IV	Do we improve performance indicator?	1. Analysis WaEx: 　1. Analyse WaEx KPR development 　2. Analyse WaEx KPI development (projects) 2. Compare WaEx KPR development with CoGr (18) KPR development	**Descr. Statistics** Average value Standard deviation Graphs	**Inferential Statistics** Goodness of fit test: Two-sample Test
6.4	H I, H II, H III, H IV	Any relationship between lean and performance indicator?	1. KPR Correlation Analysis 2. Review hypothesis		

Lean Maturity Development → Productivity Development → Review Hypotheses

WaEx = Warehouse Excellence Group (16 warehouses)
CoGr = Control Group without Lean (all in all 56, therefrom 18 that measured KPI/KPR and 38 that did not measure KPI/KPR)

We might know now if a relationship exists between lean maturity development and key performance indicators

Figure 5.3: Proof structure of the thesis

The first analysis will show if the Warehouse Excellence group implemented lean approaches and improved them. To do this, the Warehouse Excellence group will be evaluated using the BLWA. The aggregate average scores in each BLWA component will be shown for the measurement before and after the project. A deviation will be noticed when the two results are compared and this makes it possible to identify the impact of the project. If an improvement in the lean maturity is detected, the precondition for the Hypotheses I-III is established because we now know that a movement on the abscissa (see 1.2) can be shown. In order to strengthen the statement that the Warehouse Excellence group improved their lean maturity level and to be able to measure the impact of the project, hypothesis tests will be conducted to show that the first data set is significantly different from the second data set.

To establish the precondition for Hypothesis IV, the delta and deviation of the Warehouse Excellence group will be compared with the BLWA results of the control group. The comparison will be done with three different group compositions and in the following order:

- The entire control group
- The 18 warehouses that measure productivity and are the most mature in terms of lean development
- The warehouses that do not measure productivity

An additional hypothesis test will be carried out to strengthen the statement about a higher lean maturity development in the Warehouse Excellence group. The results of the hypothesis test of the Warehouse Excellence group will be compared with the test results of the control group.

The productivity development will be analyzed in the next step. First, the focus will be on productivity development from January 2010 to December 2011 at the result KPR level of the Warehouse Excellence group. Next, the productivity development from January 2010 to December 2011 will be compared at the result level for both groups. Since the lean activities began in the Warehouse Excellence group at the end of 2010, we should see a positive tendency in the group in 2011. This would support Hypothesis I. If the tendency of the Warehouse Excellence group is better than that of the control group, it would support Hypothesis IV. Additional support for Hypothesis I will involve the analysis of the productivity monitoring KPI of the Warehouse Excellence group in areas where lean techniques were implemented. Additional analysis of the lean assessment results could provide evidence about lean improvements in these areas. A comparison of the level of lean improvement with that of an increase in productivity could support Hypotheses II and III. A correlation analysis to validate Hypotheses II and III will also be done at the KPR level for the Warehouse Excellence group, the control group, and both groups together.

A final discussion on the qualitative factors and reflection on all hypotheses will complete the verification structure and finalize the research.

6 Analyzing the Lean Impact

The design of experiment and the testing structure described in chapter 5 will be used in this chapter to test the hypotheses. First, the statistical techniques that were used to analyze the data will be described and the reason for their selection will be explained. Then, an analysis of lean improvement, productivity impact, and correlation will be conducted.

6.1 Statistical Background

The descriptive statistics will be supported by graphs, tables, and characteristic values to present the generated data. Average, standard deviation, and linear regressions will be the characteristic values (Bortz and Weber, 2005; Backhaus et al., 2003) used most often in this thesis.

Unfortunately, descriptive statistics are limited by the generated data. It may not be possible to offer a statement about the population or calculate significance levels. During elections, for example, most institutes ask for a small sample size, evaluate the data, test it, and use it to generate a statement about the population. Inferential statistics are needed for such tests. These inferential statistics are divided into parametric and non-parametric tests. Parametric tests assume that the sample belongs to a population whose distribution is known. Statistical tests on samples evaluate their distribution, ascertaining, for example, if they are normally distributed. These tests are called goodness of fit tests. If a goodness of fit test cannot indicate a known distribution for the sample, the sample can be tested by a non-parametric test.

6.1.1 Choosing the Goodness of Fit Test

Lehmann and Romano (2005), Anderson and Darling (1954), Sá (2008), Cirrone et al. (2004), Duller (2008); and Janssen (2005) discussed and analyzed several goodness of fit tests. The literature review identified the following tests as the most relevant ones:

- Kolmogorov-Smirnov test
- Lilliefors test
- Chi-Squared test
- Anderson-Darling test
- Cramer-von Mises test
- Shapiro-Wilk test

Research by Janssen (2005) and Lehmann and Romano (2005) showed that the Kolmogorov-Smirnov test is more effective than the Chi-Squared test in evaluating goodness of fit. The Anderson-Darling, Cramer-von Mises, and Lilliefors tests are based on the Kolmogorov-Smirnov test. Cirrone et al. (2004) and Lehmann and Romano (2005) said that the Anderson-Darling and Cramer-von Mises tests outperform the Kolmogorov-Smirnov test because they are more sensitive indicators of the distribution. Sá (2008) compared the Kolmogorov-Smirnov test, the Shapiro-Wilk test, and the Lilliefors test. He identified the Kolmogorov-Smirnov test as the weakest of the three and the Shapiro-Wilk test as the strongest across various tested sample sizes. Razali and Wah (2011) compared the Shapiro-Wilk test, the Kolmogorov-Smirnov test, the Lilliefors test, and the Anderson-Darling test. They concluded that the Shapiro-Wilk test is the most powerful test for different distributions and sample sizes. Hence, this research uses the Shapiro-Wilk test as described in Sá (2008), Duller (2008), and Razali and Wah (2011).

If the Shapiro-Wilk test shows that the sample structures are distribution free, non-parametric tests will be used to analyze the samples.

6.1.2 Choosing the Non-Parametric Test

Janssen (2005) describes four distinguishing characteristics that must be considered before choosing a non-parametric test. These are sample size, scale level, sample quantity, and whether the samples are dependent from each other or not. Duller (2008), Janssen (2005), Sá (2008), Toutenburg et al. (2009), Genschel and Becker (2005), and Gibbons (2003) describe several non-parametric tests.

The most common tests for two independent samples with minimum ordinal scales are as follows:

- Mann-Whitney U test
- Median test
- Moses test
- Kolmogorov-Smirnov Z test
- Wald-Wolfowitz test

Toutenburg et al. (2009) and Duller (2008) describe the Mann-Whitney U, Median, and Moses tests that test specific parameters and show where the difference occurs within samples. The Mann-Whitney U and Median tests are sensitive to the location of the distribution among samples. Janssen (2005) describes the Median test as being a very general test and, hence, rather poor in its effectiveness. The Mann-Whitney test is considered to be very effective for large samples. The Moses test is sensitive to the shape of the distributions and is similar to the Wilcoxon Signed-Rank test, which will be described later. It is especially suitable when extreme reactions are expected (Duller, 2008). The advantage of the Kolmogorov-

Smirnov Z test and the Wald-Wolfowitz test is their sensitivity to the location and shape of the samples distribution. The so-called Omnibus tests that are sensitive to both criteria cannot show where the difference occurs within the samples but indicate when significant differences exist (Genschel and Becker, 2005; Gibbons, 2003; Janssen, 2005; Sá, 2008). Duller (2008) describes the Kolmogorov-Smirnov Z test as being more effective than the Wald-Wolfowitz test. In conclusion, we will use the Kolmogorov-Smirnow Z test for tests with two independent samples with minimum ordinal scales data for this thesis.

The most common tests for two dependent samples with minimum ordinal scales are (Janssen, 2005; Sá, 2008; Duller, 2008; Genschel and Becker, 2005) as follows:

- Wilcoxon Signed-Rank test
- Sign test

The Wilcoxon Signed-Rank test considers the magnitude of the difference between sample parameters. The Sign test does not, which makes the Wilcoxon Signed-Rank test more effective (Sá, 2008). This more effective aspect is also the reason why this test will be chosen for this thesis for tests with two dependent samples with minimum ordinal scales data.

All of the described statistical tests are conducted using the IBM SPSS Statistics Version 20 software. SPSS calculates significance level for each test. Clauß et al. (1994) and Bol (2003) offer the following interpretation for the calculated significance levels:

- Significance level ≤ 0.001 = high significance level
- Significance level > 0.001 and ≤ 0.010 = very significant level
- Significance level > 0.010 and ≤ 0.050 = significance level
- Significance level > 0.050 and ≤ 0.100 = low significance level
- Not significant if above > 0.100

6.2 Analysis of Lean Maturity Development

The results of the BLWA are presented in this section. The development of the lean maturity of the Warehouse Excellence group and the control group are shown by comparing the results of the first and second assessment.

6.2.1 Lean Maturity Development of the Warehouse Excellence Group

Descriptive Analysis

Chapter 4 described the development of a maturity assessment for measuring the leanness of a warehouse and section 5.1 described the project, Warehouse Excellence, while focusing on the 16 warehouses in the observation group. The Bosch Logistics Warehouse Assessment evaluated the 16 warehouses from the end of 2010 to the end of 2011 / beginning of the year 2012.

The accumulated results are highlighted in the figures 6.1, 6.2, 6.3, 6.4 and the tables 6.1, 6.2.

Table 6.3 provides figures for the assessment results over two different years. There is a high level of improvement in the overall average points as well as in the points focusing on the main lean components System-CIP and Point-CIP, which were emphasized in the Warehouse Excellence project. Additionally, the coefficient of variation shows that the spread of the function narrows.

Table 6.3 shows that the total average accumulated score improves by about 60 points, an increase of 84%. D.1 shows that each warehouse improves its entire system. Moreover, the variation coefficient

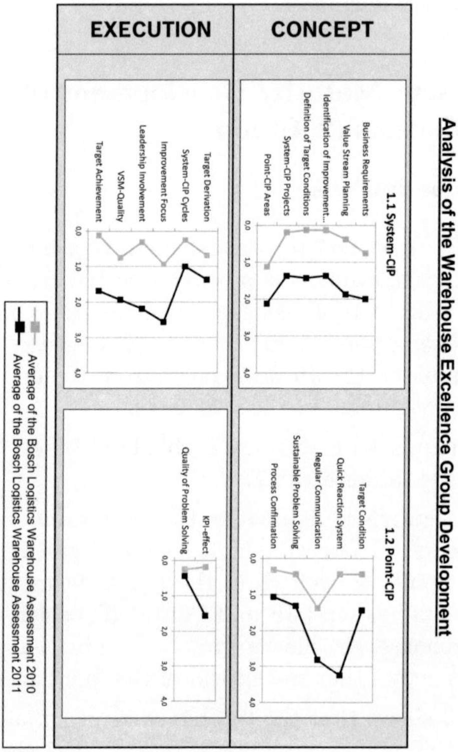

Figure 6.1: BLWA results: Warehouse Excellence group 2010 vs. Warehouse Excellence group 2011 (part 1)

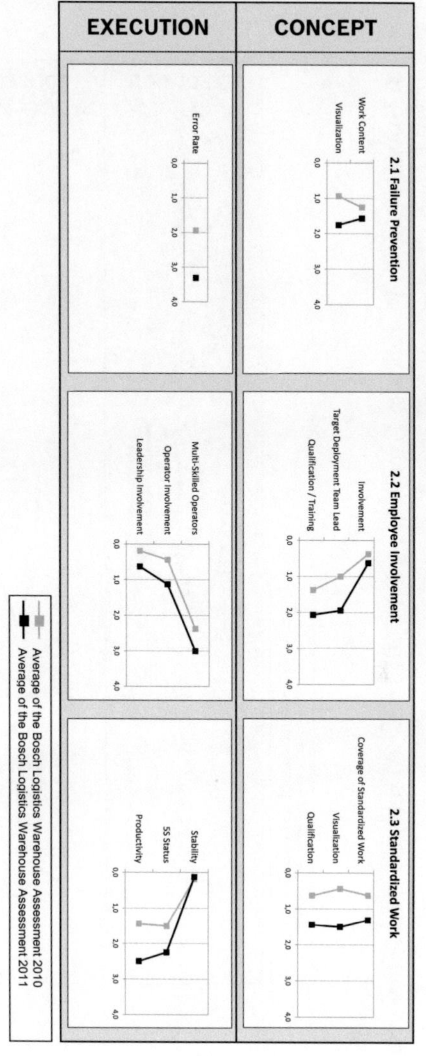

Figure 6.2: BLWA results: Warehouse Excellence group 2010 vs. Warehouse Excellence group 2011 (part 2)

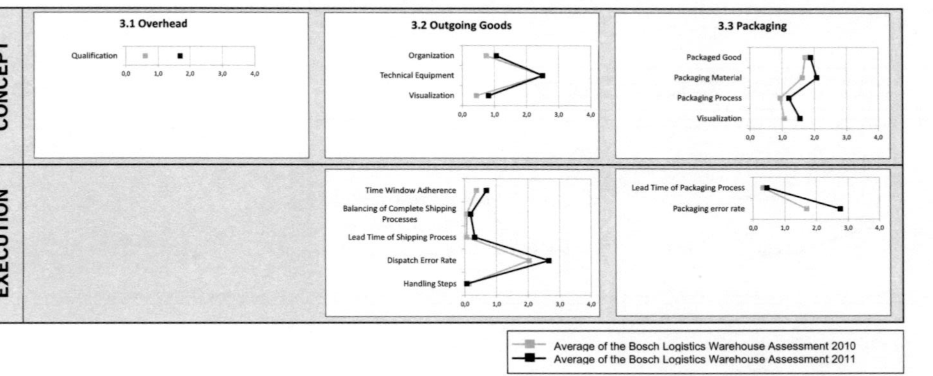

Figure 6.3: BLWA results: Warehouse Excellence group 2010 vs. Warehouse Excellence group 2011 (part 3)

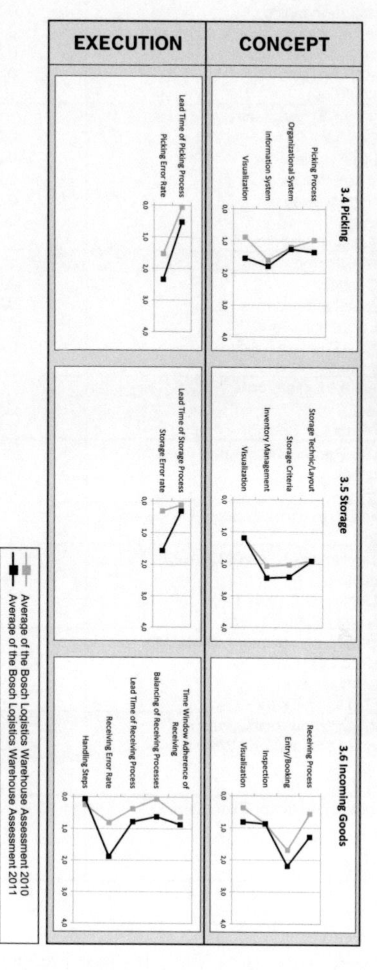

Figure 6.4: BLWA results: Warehouse Excellence group 2010 vs. Warehouse Excellence group 2011 (part 4)

Category	WaEx 2010	WaEx 2011
1.1 System-CIP Concept		
Business Requirements	0.750	2.000
Value Stream Planning	0.375	1.875
Identification of Improvement Activities	0.125	1.375
Definition of Target Conditions	0.125	1.438
System-CIP Projects	0.188	1.375
Point-CIP Areas	1.125	2.125
1.1 System-CIP Execution		
Target Derivation	0.688	1.375
System-CIP Cycles	0.250	1.000
Improvement Focus	0.938	2.563
Leadership Involvement	0.313	2.188
VSM-Quality	0.750	1.938
Target Achievement	0.125	1.688
1.2 Point-CIP Concept		
Target Condition	0.438	1.438
Quick Reaction System	0.438	3.250
Regular Communication	1.375	2.813
Sustainable Problem Solving	0.438	1.313
Process Confirmation	0.313	1.063
1.2 Point-CIP Execution		
KPI-Effect	0.188	1.563
Quality of Problem Solving	0.250	0.438
2.1 Failure Prevention System Concept		
Work Content	1.250	1.563
Visualization	0.938	1.750
2.1 Failure Prevention System Execution		
Error Rate	1.938	3.313
2.2 Employee Involvement Concept		
Involvement	0.375	0.625
Target Deployment Team Lead	1.000	1.938
Qualification / Training	1.375	2.063
2.2 Employee Involvement Execution		
Multi-Skilled Operators	2.375	3.000
Operator Involvement	0.438	1.125
Leadership Involvement	0.188	0.625
2.3 Standardized Work Concept		
Coverage of Standardized Work	0.625	1.313
Visualization	0.438	1.500
Qualification	0.625	1.438
2.3 Standardized Work Execution		
Stability	0.188	0.125
5S Status	1.500	2.250
Productivity	1.438	2.500

Table 6.1: BLWA results: Warehouse Excellence group 2010 vs. Warehouse Excellence group 2011 average scores (part 1)

Category	WaEx 2010	WaEx 2011
3.1 Overhead Concept		
Qualification	0.600	1.688
3.2 Outgoing Goods Concept		
Organization	0.750	1.063
Technical Equipment	2.500	2.500
Visualization	0.438	0.813
3.2 Outgoing Goods Execution		
Time Window Adherence	0.375	0.688
Balancing of Complete Shipping Processes	0.063	0.188
Lead Time of Shipping Process	0.063	0.313
Dispatch Error Rate	2.031	2.656
Handling Steps	0.125	0.063
3.3 Packaging Concept		
Packaged Good	1.719	1.885
Packaging Material	1.625	2.073
Packaging Process	0.938	1.208
Visualization	1.063	1.552
3.3 Packaging Execution		
Lead Time of Packaging Process	0.313	0.438
Packaging Error Rate	1.688	2.750
3.4 Picking Concept		
Picking Process	0.969	1.344
Organizational System	1.188	1.250
Information System	1.594	1.781
Visualization	0.875	1.531
3.4 Picking Execution		
Lead Time of Picking Process	0.063	0.531
Picking Error Rate	1.531	2.344
3.5 Storage Concept		
Storage Technic/Layout	1.906	1.906
Storage Criteria	2.031	2.406
Inventory Management	2.063	2.438
Visualization	1.188	1.188
3.5 Storage Execution		
Lead Time of Storage Process	0.125	0.313
Storage Error Rate	0.313	1.563
3.6 Incoming Goods Concept		
Receiving Process	0.563	1.281
Entry/Booking	1.688	2.188
Inspection	0.867	0.867
Visualization	0.375	0.813
3.6 Incoming Goods Execution		
Time Window Adherence of Receiving	0.625	0.875
Balancing of Receiving Processes	0.063	0.625
Lead Time of Receiving Process	0.375	0.781
Receiving Error Rate	0.813	1.875
Handling Steps	0.250	0.063

Table 6.2: BLWA results: Warehouse Excellence group 2010 vs. Warehouse Excellence group 2011 average scores (part 2)

	WaEx 2010	WaEx 2011	difference
Average Points in all assessment categories	57.62	105.77	+83.56 %
Coefficient of variance by all assessment categories	47.22 %	30.31 %	
Average Points in System- & Point-CIP categories	9.18	32.81	+257.14 %
Coefficient of variance by Points in System- & Point-CIP categories	88.26 %	28.35 %	

Table 6.3: BLWA results: total points

becomes smaller, which indicates a more aligned group. This means that the improvement of the average score is not just influenced by single warehouses that improved greatly while all other warehouses did not improve. It is more a sign that the entire group reached a specific positive development.

Focusing solely on the results of the System-CIP and Point-CIP, the average accumulated score rises by 257%. This increase is larger than the improvement to the total score. This also represents the focus and goals of the project. Appendix D.1 shows that before the Warehouse Excellence project only a few warehouses had established a continuous improvement cycle from process design to target condition implementation and up to process stabilization. By reaching all of the milestones, all warehouses now have a continuous improvement cycle with different maturity levels. These are also represented in the scores shown in Appendix D.1.

The System-CIP section consists of six components. The three components – identification of improvements, definition of target conditions, and System-CIP projects – are necessary for implementing specific actions. These three components show a gap in 2010 as well as in 2011 compared to other System-CIP components. The average score of the three components is 0.15 in 2010. For the others, this is 0.75 in 2010. In the year 2011, the three components had an average of 1.4, the others had 2.0. This gap narrows (viewed relatively) slightly but is still visible.

This also reflects the project team's experience during the Warehouse Excellence project. The actions that were defined by the warehouses were inaccurate. For example, the project plans did not include any capacity planning for the persons involved. Another example is that when a target condition was defined, the KPI needed to monitor that standard was not defined. Also, a systematic way of identifying the most important areas for improvement did not exist. Furthermore, the link between the goals and the projects was not shown with figures. Without these, a comprehensive strategy for creating and reaching medium-term/long-term goals can never be established. This was also one of the major findings during the project.

In the execution part of the System-CIP, the component leadership involvement had an average score of 0.313 in 2010 – one of the lowest. After focusing on the role of the leader in the CIP process, the score in 2011 was 2.188 points – one of the strongest in the category. This indicates that the taken measures were effective.

The Point-CIP results improved from an accumulated average score of 0.491 to 1.696 in 2011. This delta also shows that the taken measures were effective. In 2010, regular communication existed in practice but most warehouses lacked a well-defined scope for discussing KPIs and their deviations. Problem solving (0.438 points in 2010 and 1.313 points in 2011) and process confirmation (0.313 points in 2010 and in 1.063 points in 2011) were the two subjects with the lowest scores but still showed a clear improvement. The project team could confirm the assessment results with their experiences on the shop floors. None of the warehouses had ever had problem solving training with the shop floor team and root cause analysis was part of the training. Process confirmation was also lacking. The leaders often implement standards but did not follow up on them to ensure adherence.

Inferential Analysis - Goodness of Fit Test

This section analyzes if the changes in the assessment results of the Warehouse Excellence group were significant or not. If we know if the results are significant, we can estimate with a higher level of evidence if our results are random or based on the influence of the lean approach. In later sections, we will examine the following:

- The total assessment result for each warehouse of the Warehouse Excellence group
- The System-CIP and Point-CIP assessment results for each warehouse in the Warehouse Excellence group
- The System-CIP and Point-CIP assessment results for each warehouse in the control group

Before using a hypothesis test to describe the significant level of the results, it is necessary to test how the evaluated data is distributed (see section 6.1.1). If we know that the data is normally distributed or not, we can decide if parametric or non-parametric hypothesis tests can be used. To acquire a high level of evidence, the goodness of fit test will be done on each data set that will be tested later.

For the Shapiro-Wilk Test, we will assume the following hypotheses:

H_0: *The data set is normally distributed -> "We can use parametric two sample tests with a very high test power"*

H_1: *The data set is not normally distributed -> "We have to use non-parametric two sample tests with a good test power"*

Table 6.4 shows the results of the Shapiro-Wilk test for the total assessment result for each warehouse in the Warehouse Excellence group. Each line represents one warehouse. The abbreviation WaEx in the row stands for Warehouse Excellence group. The numbers 10 and 11 represent the data periods that are evaluated together. The first set of data is from the end of 2010 (10) and the second set is

	Shapiro-Wilk		
	Stat-istic	df	Signi-ficance
WaEx 10 & 11 W1	.699	138	.000
WaEx 10 & 11 W2	.854	138	.000
WaEx 10 & 11 W3	.789	138	.000
WaEx 10 & 11 W4	.807	138	.000
WaEx 10 & 11 W5	.758	138	.000
WaEx 10 & 11 W6	.891	138	.000
WaEx 10 & 11 W7	.769	138	.000
WaEx 10 & 11 W8	.785	138	.000
WaEx 10 & 11 W9	.855	138	.000
WaEx 10 & 11 W10	.804	138	.000
WaEx 10 & 11 W11	.879	138	.000
WaEx 10 & 11 W12	.728	138	.000
WaEx 10 & 11 W13	.773	138	.000
WaEx 10 & 11 W14	.782	138	.000
WaEx 10 & 11 W15	.854	138	.000
WaEx 10 & 11 W16	.771	138	.000

Table 6.4: Shapiro-Wilk test for the total assessment results of the Warehouse Excellence group

taken at the end of 2011 (11). The calculated significance level is below 0.001%. This means that we can reject H_0 with a high level of significance and assume that H_1 is valid. This means that no warehouse in the Warehouse Excellence group has assessment results that are normally distributed and this was expected. In conclusion, non-parametric tests should be used for the evaluation of the total assessment results of the Warehouse Excellence group.

The Hypotheses H_0 and H_1 can also be applied for the Shapiro-Wilk test in which the System-CIP and Point-CIP assessment results for each warehouse of the Warehouse Excellence group are analyzed.

Table 6.5 shows the result of testing the System-CIP and Point-CIP assessment results of the warehouses in the Warehouse Excellence group. The terminology in this table is the same that is used in the Shapiro-Wilk test table except that the letters S and P are added, which stand for System-CIP and Point-CIP. All of the warehouses tested below the significance level of 0.001% in this test as well. This means that H_0 can be rejected with a high level of significance and it can be assumed that H_1 is valid. This means we can assume that the System-CIP and Point-CIP assessment results of the warehouses in the Warehouse Excellence group are not normally distributed.

The System-CIP and Point-CIP assessment results for each warehouse in the control group were then tested to identify the distribution of the data. The hypotheses H_0 and H_1 were also applied for this test.

Table 6.6 shows the results of the testing of the System-CIP and Point-CIP assessment of the control group warehouses. Each line represents the results of one warehouse. CoGr indicates that the warehouse is from the control group. The range C1 to C56 represents the 56 warehouses in the control group. The 10 and 11 represents the time frame when the data was collected. The warehouses C9, C11, and C13 to C26 did not reach any maturity level and did not make any progress during the course of the project. They are not identified separately because no distribution exists. Interpreting the data leads to the fact that H_0 can be rejected with a similar high

	Shapiro-Wilk		
	Statistic	df	Significance
WaEx 10 & 11 S+P W1	,680	38	,000
WaEx 10 & 11 S+P W2	,869	38	,000
WaEx 10 & 11 S+P W3	,767	38	,000
WaEx 10 & 11 S+P W4	,794	38	,000
WaEx 10 & 11 S+P W5	,771	38	,000
WaEx 10 & 11 S+P W6	,853	38	,000
WaEx 10 & 11 S+P W7	,822	38	,000
WaEx 10 & 11 S+P W8	,723	38	,000
WaEx 10 & 11 S+P W9	,825	38	,000
WaEx 10 & 11 S+P W10	,785	38	,000
WaEx 10 & 11 S+P W11	,856	38	,000
WaEx 10 & 11 S+P W12	,729	38	,000
WaEx 10 & 11 S+P W13	,813	38	,000
WaEx 10 & 11 S+P W14	,684	38	,000
WaEx 10 & 11 S+P W15	,825	38	,000
WaEx 10 & 11 S+P W16	,770	38	,000

Table 6.5: Shapiro-Wilk test for the System-CIP and Point-CIP assessment results of the warehouses in the Warehouse Excellence group

	Shapiro-Wilk				Shapiro-Wilk		
	Statistic	df	Significance		Statistic	df	Significance
CoGr 10 & 11 C1	,810	38	,000	CoGr 10 & 11 C37	,737	38	,000
CoGr 10 & 11 C2	,436	38	,000	CoGr 10 & 11 C38	,803	38	,000
CoGr 10 & 11 C3	,622	38	,000	CoGr 10 & 11 C39	,763	38	,000
CoGr 10 & 11 C4	,861	38	,000	CoGr 10 & 11 C40	,701	38	,000
CoGr 10 & 11 C5	,819	38	,000	CoGr 10 & 11 C41	,750	38	,000
CoGr 10 & 11 C6	,742	38	,000	CoGr 10 & 11 C42	,731	38	,000
CoGr 10 & 11 C7	,667	38	,000	CoGr 10 & 11 C43	,752	38	,000
CoGr 10 & 11 C8	,522	38	,000	CoGr 10 & 11 C44	,751	38	,000
CoGr 10 & 11 C10	,516	38	,000	CoGr 10 & 11 C45	,701	38	,000
CoGr 10 & 11 C12	,468	38	,000	CoGr 10 & 11 C46	,509	38	,000
CoGr 10 & 11 C27	,680	38	,000	CoGr 10 & 11 C47	,663	38	,000
CoGr 10 & 11 C28	,574	38	,000	CoGr 10 & 11 C48	,787	38	,000
CoGr 10 & 11 C29	,515	38	,000	CoGr 10 & 11 C49	,720	38	,000
CoGr 10 & 11 C30	,684	38	,000	CoGr 10 & 11 C50	,152	38	,000
CoGr 10 & 11 C31	,404	38	,000	CoGr 10 & 11 C51	,836	38	,000
CoGr 10 & 11 C32	,500	38	,000	CoGr 10 & 11 C52	,152	38	,000
CoGr 10 & 11 C33	,780	38	,000	CoGr 10 & 11 C53	,498	38	,000
CoGr 10 & 11 C34	,665	38	,000	CoGr 10 & 11 C54	,325	38	,000
CoGr 10 & 11 C35	,471	38	,000	CoGr 10 & 11 C55	,826	38	,000
CoGr 10 & 11 C36	,748	38	,000	CoGr 10 & 11 C56	,599	38	,000

Table 6.6: Shapirot-Wilk test for the assessment results of the control group

significance and the assumption can be made that the lean maturity results are not normally distributed.

The result of the Shapiro-Wilk test shows that the data sets of WaEx, WaEx S+P, and CoGr respectively are not normally distributed. This needed to be known before choosing the right test (parametric or non-parametric two sample test) to analyze if the taken measures of the Warehouse Excellence project did have an impact on the maturity level of the participating warehouses.

Inferential Analysis - Wilcoxon Signed-Rank Test for the Warehouse Excellence Group

We know now with a high significance level that the distribution of the maturity development of the warehouses is not normally distributed. This means non-parametric hypothesis tests will help us to determine if the development of the assessment results of each warehouse from 2010 to 2011 is random or not. To determine which non-parametric test has to be chosen, we also need to know if the data set from 2010 and the data set from 2011 is dependent on or independent from each other.

Brosius (2011, p. 888) mentions that samples are dependent if there is a before and after comparison which is this case here. For this reason, the non-parametric two-sample dependency test that was chosen in chapter 6.1.2 will be used. The following hypotheses have been defined for the Wilcoxon Signed-Rank test:

H_0: *The samples n1 and n2 are from the same population -> "The warehouse did not improve their lean maturity"*

H_1: *The samples n1 and n2 are not from the same population -> "The warehouse did improve their lean maturity"*

The sample n1 is the data set from the assessment results for the year 2010. The sample n2 is from the assessment results for the year 2011.

Table 6.7 shows the ranking table of the Wilcoxon Signed-Rank test total assessment results for each warehouse in the Warehouse Excellence group. In this analysis, the results obtained from the beginning of the project were compared with the results obtained at the end of the project. A negative ranking means that the warehouse had a negative development in an assessment component. A positive ranking means that development in a component was positive. Similarly, a tie means there has been no change in the specific component. Seventy components were considered in total. However, the warehouses had an overall negative ranking in 1.4% of the cases. In 57.3% of the cases, there were ties and there was a positive ranking in 41.3% of the cases. All warehouses had more positive rankings than negative ones. This means that the taken measures, during the Warehouse Excellence project, could be seen as a positive lean maturity development.

Table 6.8 shows the test statistics for the Wilcoxon Signed-Rank test. Each box represents the before and after comparison in a particular warehouse. 15 of the 16 warehouses shows with high significance and one warehouse with significance that H_0 can be rejected. This means that it can be assumed that there is a significant difference between the samples. In other words, the results from the year 2010 are so different from the year 2011 that it cannot be random. This indicates an overall high significance that the taken measures from the Warehouse Excellence project influenced the warehouses and could be seen by the assessments results.

Analogous to the Wilcoxon Signed-Rank test, the analysis was conducted with the data set from the System-CIP and Point-CIP assessment results of the Warehouse Excellence group.

Table 6.9 shows the ranking table of the Wilcoxon Signed-Rank test for the assessment results of the System-CIP and Point-CIP results of the Warehouse Excellence group. The warehouses show negative

		N	Mean Rank	Sum of Ranks			N	Mean Rank	Sum of Ranks
WaEx W1 2011 - WaEx W1 2010	Negative Ranks	0[a]	0.00	0.00	WaEx W9 2011 - WaEx W9 2010	Negative Ranks	4[y]	21.63	86.50
	Positive Ranks	33[b]	17.00	561.00		Positive Ranks	36[z]	20.38	733.50
	Ties	37[c]				Ties	30[aa]		
	Total	70				Total	70		
WaEx W2 2011 - WaEx W2 2010	Negative Ranks	1[d]	35.50	35.50	WaEx W10 2011 - WaEx W10 2010	Negative Ranks	3[ab]	12.33	37.00
	Positive Ranks	36[e]	18.54	667.50		Positive Ranks	24[ac]	14.21	341.00
	Ties	33[f]				Ties	43[ad]		
	Total	70				Total	70		
WaEx W3 2011 - WaEx W3 2010	Negative Ranks	2[g]	14.75	29.50	WaEx W11 2011 - WaEx W11 2010	Negative Ranks	3[ae]	11.83	35.50
	Positive Ranks	19[h]	10.61	201.50		Positive Ranks	20[af]	12.03	240.50
	Ties	49[i]				Ties	46[ag]		
	Total	70				Total	69		
WaEx W4 2011 - WaEx W4 2010	Negative Ranks	1[j]	11.00	11.00	WaEx W12 2011 - WaEx W12 2010	Negative Ranks	1[ah]	6.50	6.50
	Positive Ranks	41[k]	21.76	892.00		Positive Ranks	19[ai]	10.71	203.50
	Ties	28[l]				Ties	50[aj]		
	Total	70				Total	70		
WaEx W5 2011 - WaEx W5 2010	Negative Ranks	0[m]	0.00	0.00	WaEx W13 2011 - WaEx W13 2010	Negative Ranks	0[ak]	0.00	0.00
	Positive Ranks	34[n]	17.50	595.00		Positive Ranks	31[al]	16.00	496.00
	Ties	36[o]				Ties	39[am]		
	Total	70				Total	70		
WaEx W6 2011 - WaEx W6 2010	Negative Ranks	1[p]	21.50	21.50	WaEx W14 2011 - WaEx W14 2010	Negative Ranks	0[an]	0.00	0.00
	Positive Ranks	32[q]	16.86	539.50		Positive Ranks	19[ao]	10.00	190.00
	Ties	37[r]				Ties	51[ap]		
	Total	70				Total	70		
WaEx W7 2011 - WaEx W7 2010	Negative Ranks	0[s]	0.00	0.00	WaEx W15 2011 - WaEx W15 2010	Negative Ranks	0[aq]	0.00	0.00
	Positive Ranks	27[t]	14.00	378.00		Positive Ranks	28[ar]	14.50	406.00
	Ties	43[u]				Ties	42[as]		
	Total	70				Total	70		
WaEx W8 2011 - WaEx W8 2010	Negative Ranks	0[v]	0.00	0.00	WaEx W16 2011 - WaEx W16 2010	Negative Ranks	0[at]	0.00	0.00
	Positive Ranks	24[w]	12.50	300.00		Positive Ranks	38[au]	19.50	741.00
	Ties	46[x]				Ties	32[av]		
	Total	70				Total	70		

a. WaEx W1 2011 < WaEx W1 2010
b. WaEx W1 2011 > WaEx W1 2010
c. WaEx W1 2011 = WaEx W1 2010

● ● ●

between "d" and "as" similar

at. WaEx W16 2011 < WaEx W16 2010
au. WaEx W16 2011 > WaEx W16 2010
av. WaEx W16 2011 = WaEx W16 2010

Table 6.7: Wilcoxon Rank table for the assessment results of the warehouses in the Warehouse Excellence group

Test Statistics[a]

	WaEx W1 2011 - WaEx W1 2010	WaEx W2 2011 - WaEx W2 2010	WaEx W3 2011 - WaEx W3 2010	WaEx W4 2011 - WaEx W4 2010
Z	$-5{,}232^b$	$-4{,}899^b$	$-3{,}236^b$	$-5{,}610^b$
Asymp. Sig. (2-tailed)	.000	.000	.001	.000

	WaEx W5 2011 - WaEx W5 2010	WaEx W6 2011 - WaEx W6 2010	WaEx W7 2011 - WaEx W7 2010	WaEx W8 2011 - WaEx W8 2010
Z	$-5{,}157^b$	$-4{,}710^b$	$-4{,}724^b$	$-4{,}485^b$
Asymp. Sig. (2-tailed)	.000	.000	.000	.000

	WaEx W9 2011 - WaEx W9 2010	WaEx W10 2011 - WaEx W10 2010	WaEx W11 2011 - WaEx W11 2010	WaEx W12 2011 - WaEx W12 2010
Z	$-4{,}396^b$	$-3{,}754^b$	$-3{,}171^b$	$-3{,}780^b$
Asymp. Sig. (2-tailed)	.000	.000	.002	.000

	WaEx W13 2011 - WaEx W13 2010	WaEx W14 2011 - WaEx W14 2010	WaEx W15 2011 - WaEx W15 2010	WaEx W16 2011 - WaEx W16 2010
Z	$-4{,}944^b$	$-4{,}014^b$	$-4{,}671^b$	$-5{,}443^b$
Asymp. Sig. (2-tailed)	.000	.000	.000	.000

a. Wilcoxon Test

b. Based on negative ranks.

Table 6.8: Wilcoxon test statistics for the assessment results of the warehouses in the Warehouse Excellence group

		N	Mean Rank	Sum of Ranks			N	Mean Rank	Sum of Ranks
WaEx W1 2011 S+P - WaEx W1 2010 S+P	Negative Ranks	0[a]	0.00	0.00	WaEx W9 2011 S+P - WaEx W9 2010 S+P	Negative Ranks	0[y]	0.00	0.00
	Positive Ranks	16[b]	8.50	136.00		Positive Ranks	15[z]	8.00	120.00
	Ties	3[c]				Ties	4[aa]		
	Total	19				Total	19		
WaEx W2 2011 S+P - WaEx W2 2010 S+P	Negative Ranks	0[d]	0.00	0.00	WaEx W10 2011 S+P - WaEx W10 2010 S+P	Negative Ranks	0[ab]	0.00	0.00
	Positive Ranks	15[e]	8.00	120.00		Positive Ranks	14[ac]	7.50	105.00
	Ties	4[f]				Ties	5[ad]		
	Total	19				Total	19		
WaEx W3 2011 S+P - WaEx W3 2010 S+P	Negative Ranks	0[g]	0.00	0.00	WaEx W11 2011 S+P - WaEx W11 2010 S+P	Negative Ranks	3[ae]	6.33	19.00
	Positive Ranks	9[h]	5.00	45.00		Positive Ranks	7[af]	5.14	36.00
	Ties	10[i]				Ties	9[ag]		
	Total	19				Total	19		
WaEx W4 2011 S+P - WaEx W4 2010 S+P	Negative Ranks	0[j]	0.00	0.00	WaEx W12 2011 S+P - WaEx W12 2010 S+P	Negative Ranks	0[ah]	0.00	0.00
	Positive Ranks	15[k]	8.00	120.00		Positive Ranks	14[ai]	7.50	105.00
	Ties	4[l]				Ties	5[aj]		
	Total	19				Total	19		
WaEx W5 2011 S+P - WaEx W5 2010 S+P	Negative Ranks	0[m]	0.00	0.00	WaEx W13 2011 S+P - WaEx W13 2010 S+P	Negative Ranks	0[ak]	0.00	0.00
	Positive Ranks	17[n]	9.00	153.00		Positive Ranks	15[al]	8.00	120.00
	Ties	2[o]				Ties	4[am]		
	Total	19				Total	19		
WaEx W6 2011 S+P - WaEx W6 2010 S+P	Negative Ranks	0[p]	0.00	0.00	WaEx W14 2011 S+P - WaEx W14 2010 S+P	Negative Ranks	0[an]	0.00	0.00
	Positive Ranks	12[q]	6.50	78.00		Positive Ranks	7[ao]	4.00	28.00
	Ties	7[r]				Ties	12[ap]		
	Total	19				Total	19		
WaEx W7 2011 S+P - WaEx W7 2010 S+P	Negative Ranks	0[s]	0.00	0.00	WaEx W15 2011 S+P - WaEx W15 2010 S+P	Negative Ranks	0[aq]	0.00	0.00
	Positive Ranks	12[t]	6.50	78.00		Positive Ranks	13[ar]	7.00	91.00
	Ties	7[u]				Ties	6[as]		
	Total	19				Total	19		
WaEx W8 2011 S+P - WaEx W8 2010 S+P	Negative Ranks	0[v]	0.00	0.00	WaEx W16 2011 S+P - WaEx W16 2010 S+P	Negative Ranks	0[at]	0.00	0.00
	Positive Ranks	14[w]	7.50	105.00		Positive Ranks	17[au]	9.00	153.00
	Ties	5[x]				Ties	2[av]		
	Total	19				Total	19		

a. WaEx W1 2011 < WaEx W1 2010
b. WaEx W1 2011 > WaEx W1 2010
c. WaEx W1 2011 = WaEx W1 2010

● ● ●
between "d" and
"as" similar

at. WaEx W16 2011 < WaEx W16 2010
au. WaEx W16 2011 > WaEx W16 2010
av. WaEx W16 2011 = WaEx W16 2010

Table 6.9: Wilcoxon Rank table for the assessment results for the System-CIP and Point-CIP of the warehouses in the Warehouse Excellence group

rankings in 1% of the cases, positive rankings in 69.7 % of the cases, and a tie in 29.3% of the cases. Similar to the earlier results, a positive trend is shown for the taken measures that were implemented during the Warehouse Excellence project. The higher positive trend compared to the results of the test that was done before with the data set of the total assessment underlines the focus of the project: the implementation of a systematic and analytical continuous improvement cycle with the System-CIP and Point-CIP approach.

Table 6.10 shows the test statistics of the Wilcoxon-Signed-Rank test for the assessment results of the System-CIP and Point-CIP results of the Warehouse Excellence group. In 11 warehouses, there is with a high significance level that H_0 can be rejected. In four warehouses, the results were very significant and in one warehouse the significance level of 0.374 is too low to reject H_0. The reason for the less strong significance level, compared to the first Wilcoxon-Signed-Rank test statistics, is that the sample size is smaller. Single negative cases have a stronger effect on the results because the lower total amount of ties does not absorb the single negative impact. In turn, the other results are very strong despite the small sample size.

Inferential Analysis - Wilcoxon Signed-Rank test for the Control Group

In order to evaluate the results of the Wilcoxon Signed-Rank test for the Warehouse Excellence group, we also analyzed each warehouse in the control group. Tables 6.11 and 6.12 give the ranking tables of the Wilcoxon-Signed-Rank test for the assessment results of the System-CIP and Point-CIP components of the control group. The warehouses had negative rankings in 9.3% of the cases, positive rankings in 18.9% of the cases, and ties in 71.8% of the cases. This represents a moderately positive trend.

Seventeen warehouses have ties in all components. This indicates that the issue was not totally addressed in the warehouses. Excluding the warehouses with the 19 ties, the results are also moderately

Test Statistics[a]

	WaEx W1 2011 S+P - WaEx W1 2010 S+P	WaEx W2 2011 S+P - WaEx W2 2010 S+P	WaEx W3 2011 S+P - WaEx W3 2010 S+P	WaEx W4 2011 S+P - WaEx W4 2010 S+P
Z	-3,656[b]	-3,535[b]	-2,810[b]	-3,446[b]
Asymp. Sig. (2-tailed)	.000	.000	.005	.001

	WaEx W5 2011 S+P - WaEx W5 2010 S+P	WaEx W6 2011 S+P - WaEx W6 2010 S+P	WaEx W7 2011 S+P - WaEx W7 2010 S+P	WaEx W8 2011 S+P - WaEx W8 2010 S+P
Z	-3,660[b]	-3,115[b]	-3,126[b]	-3,407[b]
Asymp. Sig. (2-tailed)	.000	.002	.002	.001

	WaEx W9 2011 S+P - WaEx W9 2010 S+P	WaEx W10 2011 S+P - WaEx W10 2010 S+P	WaEx W11 2011 S+P - WaEx W11 2010 S+P	WaEx W12 2011 S+P - WaEx W12 2010 S+P
Z	-3,449[b]	-3,360[b]	-,890[b]	-3,370[b]
Asymp. Sig. (2-tailed)	.001	.001	.374	.001

	WaEx W13 2011 S+P - WaEx W13 2010 S+P	WaEx W14 2011 S+P - WaEx W14 2010 S+P	WaEx W15 2011 S+P - WaEx W15 2010 S+P	WaEx W16 2011 S+P - WaEx W16 2010 S+P
Z	-3,501[b]	-2,456[b]	-3,235[b]	-3,663[b]
Asymp. Sig. (2-tailed)	.000	.014	.001	.000

a. Wilcoxon-Test

b. Based on negative ranks.

Table 6.10: Wilcoxon test statistics for the assessment results for the System-CIP and Point-CIP of the warehouses in the Warehouse Excellence group

Warehouses C1–C10

		N	Mean Rank	Sum of Ranks
CoGr 2011 C1 - CoGr 2010 C1	Negative Ranks	5[aw]	6,20	31,00
	Positive Ranks	7[ax]	6,71	47,00
	Ties	7[ay]		
	Total	19		
CoGr 2011 C2 - CoGr 2010 C2	Negative Ranks	4[az]	2,50	10,00
	Positive Ranks	0[ba]	0,00	0,00
	Ties	15[bb]		
	Total	19		
CoGr 2011 C3 - CoGr 2010 C3	Negative Ranks	2[bc]	5,00	10,00
	Positive Ranks	6[bd]	4,33	26,00
	Ties	11[be]		
	Total	19		
CoGr 2011 C4 - CoGr 2010 C4	Negative Ranks	1[bf]	4,50	4,50
	Positive Ranks	15[bg]	8,77	131,50
	Ties	3[bh]		
	Total	19		
CoGr 2011 C5 - CoGr 2010 C5	Negative Ranks	3[bi]	6,17	18,50
	Positive Ranks	16[bj]	10,72	171,50
	Ties	0[bk]		
	Total	19		
CoGr 2011 C6 - CoGr 2010 C6	Negative Ranks	2[bl]	4,75	9,50
	Positive Ranks	6[bm]	4,42	26,50
	Ties	11[bn]		
	Total	19		
CoGr 2011 C7 - CoGr 2010 C7	Negative Ranks	0[bo]	0,00	0,00
	Positive Ranks	0[bp]	0,00	0,00
	Ties	19[bq]		
	Total	19		
CoGr 2011 C8 - CoGr 2010 C8	Negative Ranks	1[br]	3,50	3,50
	Positive Ranks	4[bs]	2,88	11,50
	Ties	14[bt]		
	Total	19		
CoGr 2011 C9 - CoGr 2010 C9	Negative Ranks	0[bu]	0,00	0,00
	Positive Ranks	0[bv]	0,00	0,00
	Ties	19[bw]		
	Total	19		
CoGr 2011 C10 - CoGr 2010 C10	Negative Ranks	0[bx]	0,00	0,00
	Positive Ranks	9[by]	5,00	45,00
	Ties	10[bz]		
	Total	19		

Warehouses C11–C20

		N	Mean Rank	Sum of Ranks
CoGr 2011 C11 - CoGr 2010 C11	Negative Ranks	0[ca]	0,00	0,00
	Positive Ranks	0[cb]	0,00	0,00
	Ties	19[cc]		
	Total	19		
CoGr 2011 C12 - CoGr 2010 C12	Negative Ranks	1[cd]	3,50	3,50
	Positive Ranks	3[ce]	2,17	6,50
	Ties	15[cf]		
	Total	19		
CoGr 2011 C13 - CoGr 2010 C13	Negative Ranks	0[cg]	0,00	0,00
	Positive Ranks	0[ch]	0,00	0,00
	Ties	19[ci]		
	Total	19		
CoGr 2011 C14 - CoGr 2010 C14	Negative Ranks	0[cj]	0,00	0,00
	Positive Ranks	0[ck]	0,00	0,00
	Ties	19[cl]		
	Total	19		
CoGr 2011 C15 - CoGr 2010 C15	Negative Ranks	0[cm]	0,00	0,00
	Positive Ranks	0[cn]	0,00	0,00
	Ties	19[co]		
	Total	19		
CoGr 2011 C16 - CoGr 2010 C16	Negative Ranks	0[cp]	0,00	0,00
	Positive Ranks	0[cq]	0,00	0,00
	Ties	19[cr]		
	Total	19		
CoGr 2011 C17 - CoGr 2010 C17	Negative Ranks	0[cs]	0,00	0,00
	Positive Ranks	0[ct]	0,00	0,00
	Ties	19[cu]		
	Total	19		
CoGr 2011 C18 - CoGr 2010 C18	Negative Ranks	0[cv]	0,00	0,00
	Positive Ranks	0[cw]	0,00	0,00
	Ties	19[cx]		
	Total	19		
CoGr 2011 C19 - CoGr 2010 C19	Negative Ranks	0[cy]	0,00	0,00
	Positive Ranks	0[cz]	0,00	0,00
	Ties	19[da]		
	Total	19		
CoGr 2011 C20 - CoGr 2010 C20	Negative Ranks	0[db]	0,00	0,00
	Positive Ranks	0[dc]	0,00	0,00
	Ties	19[dd]		
	Total	19		

Warehouses C21–C30

		N	Mean Rank	Sum of Ranks
CoGr 2011 C21 - CoGr 2010 C21	Negative Ranks	0[de]	0,00	0,00
	Positive Ranks	0[df]	0,00	0,00
	Ties	19[dg]		
	Total	19		
CoGr 2011 C22 - CoGr 2010 C22	Negative Ranks	0[dh]	0,00	0,00
	Positive Ranks	0[di]	0,00	0,00
	Ties	19[dj]		
	Total	19		
CoGr 2011 C23 - CoGr 2010 C23	Negative Ranks	0[dk]	0,00	0,00
	Positive Ranks	0[dl]	0,00	0,00
	Ties	19[dm]		
	Total	19		
CoGr 2011 C24 - CoGr 2010 C24	Negative Ranks	0[dn]	0,00	0,00
	Positive Ranks	0[do]	0,00	0,00
	Ties	19[dp]		
	Total	19		
CoGr 2011 C25 - CoGr 2010 C25	Negative Ranks	0[dq]	0,00	0,00
	Positive Ranks	0[dr]	0,00	0,00
	Ties	19[ds]		
	Total	19		
CoGr 2011 C26 - CoGr 2010 C26	Negative Ranks	0[dt]	0,00	0,00
	Positive Ranks	0[du]	0,00	0,00
	Ties	19[dv]		
	Total	19		
CoGr 2011 C27 - CoGr 2010 C27	Negative Ranks	8[dw]	4,63	37,00
	Positive Ranks	1[dx]	8,00	8,00
	Ties	10[dy]		
	Total	19		
CoGr 2011 C28 - CoGr 2010 C28	Negative Ranks	5[dz]	3,80	19,00
	Positive Ranks	1[ea]	2,00	2,00
	Ties	13[eb]		
	Total	19		
CoGr 2011 C29 - CoGr 2010 C29	Negative Ranks	4[ec]	3,38	13,50
	Positive Ranks	1[ed]	1,50	1,50
	Ties	14[ee]		
	Total	19		
CoGr 2011 C30 - CoGr 2010 C30	Negative Ranks	1[ef]	1,00	1,00
	Positive Ranks	0[eg]	0,00	0,00
	Ties	18[eh]		
	Total	19		

aw. CoGr 2011 C1 < CoGr 2010 C1
ax. CoGr 2011 C1 > CoGr 2010 C1
ay. CoGr 2011 C1 = CoGr 2010 C1

• • •
between "aw" and "eh" similar

ef. CoGr 2011 C30 < CoGr 2010 C30
eg. CoGr 2011 C30 > CoGr 2010 C30
eh. CoGr 2011 C30 = CoGr 2010 C30

Table 6.11: Wilcoxon Rank table for the assessment results for the System-CIP and Point-CIP of the warehouses C1-C30 in the control group

		N	Mean Rank	Sum of Ranks
CoGr 2011 C31 - CoGr 2010 C31	Negative Ranks	1[ei]	4.00	4.00
	Positive Ranks	4[ej]	2.75	11.00
	Ties	14[ek]		
	Total	19		
CoGr 2011 C32 - CoGr 2010 C32	Negative Ranks	3[el]	3.33	10.00
	Positive Ranks	3[em]	3.67	11.00
	Ties	13[en]		
	Total	19		
CoGr 2011 C33 - CoGr 2010 C33	Negative Ranks	5[eo]	6.90	34.50
	Positive Ranks	7[ep]	6.21	43.50
	Ties	7[eq]		
	Total	19		
CoGr 2011 C34 - CoGr 2010 C34	Negative Ranks	0[er]	0.00	0.00
	Positive Ranks	8[es]	4.50	36.00
	Ties	11[et]		
	Total	19		
CoGr 2011 C35 - CoGr 2010 C35	Negative Ranks	0[eu]	0.00	0.00
	Positive Ranks	1[ev]	1.00	1.00
	Ties	18[ew]		
	Total	19		
CoGr 2011 C36 - CoGr 2010 C36	Negative Ranks	5[ex]	5.90	29.50
	Positive Ranks	5[ey]	5.10	25.50
	Ties	9[ez]		
	Total	19		
CoGr 2011 C37 - CoGr 2010 C37	Negative Ranks	1[fa]	4.00	4.00
	Positive Ranks	7[fb]	4.57	32.00
	Ties	11[fc]		
	Total	19		
CoGr 2011 C38 - CoGr 2010 C38	Negative Ranks	6[fd]	3.67	22.00
	Positive Ranks	1[fe]	6.00	6.00
	Ties	12[ff]		
	Total	19		
CoGr 2011 C39 - CoGr 2010 C39	Negative Ranks	0[fg]	0.00	0.00
	Positive Ranks	8[fh]	4.50	36.00
	Ties	11[fi]		
	Total	19		

		N	Mean Rank	Sum of Ranks
CoGr 2011 C40 - CoGr 2010 C40	Negative Ranks	2[fj]	5.00	10.00
	Positive Ranks	7[fk]	5.00	35.00
	Ties	10[fl]		
	Total	19		
CoGr 2011 C41 - CoGr 2010 C41	Negative Ranks	3[fm]	2.67	8.00
	Positive Ranks	3[fn]	4.33	13.00
	Ties	13[fo]		
	Total	19		
CoGr 2011 C42 - CoGr 2010 C42	Negative Ranks	1[fp]	5.50	5.50
	Positive Ranks	7[fq]	4.36	30.50
	Ties	11[fr]		
	Total	19		
CoGr 2011 C43 - CoGr 2010 C43	Negative Ranks	0[fs]	0.00	0.00
	Positive Ranks	7[ft]	4.00	28.00
	Ties	12[fu]		
	Total	19		
CoGr 2011 C44 - CoGr 2010 C44	Negative Ranks	3[fv]	2.50	7.50
	Positive Ranks	8[fw]	7.31	58.50
	Ties	8[fx]		
	Total	19		
CoGr 2011 C45 - CoGr 2010 C45	Negative Ranks	3[fy]	3.50	10.50
	Positive Ranks	5[fz]	5.10	25.50
	Ties	11[ga]		
	Total	19		
CoGr 2011 C46 - CoGr 2010 C46	Negative Ranks	5[gb]	3.70	18.50
	Positive Ranks	1[gc]	2.50	2.50
	Ties	13[gd]		
	Total	19		
CoGr 2011 C47 - CoGr 2010 C47	Negative Ranks	7[ge]	5.57	39.00
	Positive Ranks	2[gf]	3.00	6.00
	Ties	10[gg]		
	Total	19		
CoGr 2011 C48 - CoGr 2010 C48	Negative Ranks	1[gh]	2.00	2.00
	Positive Ranks	10[gi]	6.40	64.00
	Ties	8[gj]		
	Total	19		

		N	Mean Rank	Sum of Ranks
CoGr 2011 C49 - CoGr 2010 C49	Negative Ranks	3[gk]	5.00	15.00
	Positive Ranks	6[gl]	5.00	30.00
	Ties	10[gm]		
	Total	19		
CoGr 2011 C50 - CoGr 2010 C50	Negative Ranks	0[gn]	0.00	0.00
	Positive Ranks	1[go]	1.00	1.00
	Ties	18[gp]		
	Total	19		
CoGr 2011 C51 - CoGr 2010 C51	Negative Ranks	4[gq]	6.00	24.00
	Positive Ranks	11[gr]	8.73	96.00
	Ties	4[gs]		
	Total	19		
CoGr 2011 C52 - CoGr 2010 C52	Negative Ranks	0[gt]	0.00	0.00
	Positive Ranks	1[gu]	1.00	1.00
	Ties	18[gv]		
	Total	19		
CoGr 2011 C53 - CoGr 2010 C53	Negative Ranks	4[gw]	2.50	10.00
	Positive Ranks	0[gx]	0.00	0.00
	Ties	15[gy]		
	Total	19		
CoGr 2011 C54 - CoGr 2010 C54	Negative Ranks	0[gz]	0.00	0.00
	Positive Ranks	4[ha]	2.50	10.00
	Ties	15[hb]		
	Total	19		
CoGr 2011 C55 - CoGr 2010 C55	Negative Ranks	3[hc]	4.00	12.00
	Positive Ranks	9[hd]	7.33	66.00
	Ties	7[he]		
	Total	19		
CoGr 2011 C56 - CoGr 2010 C56	Negative Ranks	2[hf]	3.00	6.00
	Positive Ranks	6[hg]	5.00	30.00
	Ties	11[hh]		
	Total	19		

ei. CoGr 2011 C31 < CoGr 2010 C31
ej. CoGr 2011 C31 > CoGr 2010 C31
ek. CoGr 2011 C31 = CoGr 2010 C31

• • •
between "aw" and
"hh" similar

hf. CoGr 2011 C56 < CoGr 2010 C56
hg. CoGr 2011 C56 > CoGr 2010 C56
hh. CoGr 2011 C56 = CoGr 2010 C56

Table 6.12: Wilcoxon Rank table for the assessment results for the System-CIP and Point-CIP of the warehouses C31-C56 in the control group

positive: 13.4% of the cases have a negative ranking, 27.1% of the cases show a positive trend, and 59.55% of the rankings were ties.

Tables 6.13 and 6.14 show the test statistics for the Wilcoxon-Signed-Rank test for the assessment results of the System-CIP and Point-CIP results of the control group. H_0 can be rejected

- with a high significance level for one warehouse,
- with a very significance level for five warehouses,
- with significance for six warehouses and
- with a low significance level for nine warehouses.

In other words, it can be assumed, that in 21 warehouses, with minimum low significance level, the warehouses did improve their lean maturity. In the case of the other 35 warehouse, H_0 cannot be rejected and this indicates that these warehouses did not improve their lean maturity significant.

Summarized for the control group this means that just 37.5% of the warehouses did improve their lean maturity with a minimum low significance level. Compared to the Warehouse Excellence group, in which 93.75% of the warehouses showed with minimum a very significance level an improvement, this means that the control group improved absolute and relative much less.

6.2.2 The Warehouse Excellence Group versus the Control Group

In earlier sections, we analyzed the development of the Warehouse Excellence group. We also compared the test statistics results of the Warehouse Excellence group with the control group. In this section we will describe the direct comparison of the assessment results of the Warehouse Excellence group with the control group. This will be done first with the entire control group of 56 warehouses,

Test Statistics[a]

	CoGr 2011 C1 - CoGr 2010 C1	CoGr 2011 C2 - CoGr 2010 C2	CoGr 2011 C3 - CoGr 2010 C3	CoGr 2011 C4 - CoGr 2010 C4
Z	$-,644^b$	$-1,841^c$	$-1,140^b$	$-3,337^b$
Asymp. Sig. (2-tailed)	.519	.066	.254	.001

	CoGr 2011 C5 - CoGr 2010 C5	CoGr 2011 C6 - CoGr 2010 C6	CoGr 2011 C7 - CoGr 2010 C7	CoGr 2011 C8 - CoGr 2010 C8
Z	$-3,111^b$	$-1,199^b$	$,000^d$	$-1,131^b$
Asymp. Sig. (2-tailed)	.002	.230	1.000	.258

	CoGr 2011 C9 - CoGr 2010 C9	CoGr 2011 C10 - CoGr 2010 C10	CoGr 2011 C11 - CoGr 2010 C11	CoGr 2011 C12 - CoGr 2010 C12
Z	$,000^d$	$-2,751^b$	$,000^d$	$-,552^b$
Asymp. Sig. (2-tailed)	1.000	.006	1.000	.581

	CoGr 2011 C13 - CoGr 2010 C13	CoGr 2011 C14 - CoGr 2010 C14	CoGr 2011 C15 - CoGr 2010 C15	CoGr 2011 C16 - CoGr 2010 C16
Z	$,000^d$	$,000^d$	$,000^d$	$,000^d$
Asymp. Sig. (2-tailed)	1.000	1.000	1.000	1.000

	CoGr 2011 C17 - CoGr 2010 C17	CoGr 2011 C18 - CoGr 2010 C18	CoGr 2011 C19 - CoGr 2010 C19	CoGr 2011 C20 - CoGr 2010 C20
Z	$,000^d$	$,000^d$	$,000^d$	$,000^d$
Asymp. Sig. (2-tailed)	1.000	1.000	1.000	1.000

	CoGr 2011 C21 - CoGr 2010 C21	CoGr 2011 C22 - CoGr 2010 C22	CoGr 2011 C23 - CoGr 2010 C23	CoGr 2011 C24 - CoGr 2010 C24
Z	$,000^d$	$,000^d$	$,000^d$	$,000^d$
Asymp. Sig. (2-tailed)	1.000	1.000	1.000	1.000

	CoGr 2011 C25 - CoGr 2010 C25	CoGr 2011 C26 - CoGr 2010 C26	CoGr 2011 C27 - CoGr 2010 C27	CoGr 2011 C28 - CoGr 2010 C28
Z	$,000^d$	$,000^d$	$-1,780^c$	$-1,807^c$
Asymp. Sig. (2-tailed)	1.000	1.000	.075	.071

a. Wilcoxon Test

b. Based on negative ranks.

b. Based on positive ranks.

d. The sum of the negative ranks is equal the sum of the positive ranks

Table 6.13: Wilcoxon test statistics for the assessment results for the System-CIP and Point-CIP of the warehouses C1-C28 in the control group

Test Statistics[a]

	CoGr 2011 C29 - CoGr 2010 C29	CoGr 2011 C30 - CoGr 2010 C30	CoGr 2011 C31 - CoGr 2010 C31	CoGr 2011 C32 - CoGr 2010 C32
Z	$-1,633^c$	$-1,000^c$	$-,966^b$	$-,105^b$
Asymp. Sig. (2-tailed)	.102	.317	.334	.916

	CoGr 2011 C33 - CoGr 2010 C33	CoGr 2011 C34 - CoGr 2010 C34	CoGr 2011 C35 - CoGr 2010 C35	CoGr 2011 C36 - CoGr 2010 C36
Z	$-,365^b$	$-2,636^b$	$-1,000^b$	$-,207^c$
Asymp. Sig. (2-tailed)	.715	.008	.317	.836

	CoGr 2011 C37 - CoGr 2010 C37	CoGr 2011 C38 - CoGr 2010 C38	CoGr 2011 C39 - CoGr 2010 C39	CoGr 2011 C40 - CoGr 2010 C40
Z	$-2,111^b$	$-1,364^c$	$-2,588^b$	$-1,667^b$
Asymp. Sig. (2-tailed)	.035	.172	.010	.096

	CoGr 2011 C41 - CoGr 2010 C41	CoGr 2011 C42 - CoGr 2010 C42	CoGr 2011 C43 - CoGr 2010 C43	CoGr 2011 C44 - CoGr 2010 C44
Z	$-,539^b$	$-1,781^b$	$-2,401^b$	$-2,303^b$
Asymp. Sig. (2-tailed)	.590	.075	.016	.021

	CoGr 2011 C45 - CoGr 2010 C45	CoGr 2011 C46 - CoGr 2010 C46	CoGr 2011 C47 - CoGr 2010 C47	CoGr 2011 C48 - CoGr 2010 C48
Z	$-1,100^b$	$-1,725^c$	$-1,997^c$	$-2,791^b$
Asymp. Sig. (2-tailed)	.271	.084	.046	.005

	CoGr 2011 C49 - CoGr 2010 C49	CoGr 2011 C50 - CoGr 2010 C50	CoGr 2011 C51 - CoGr 2010 C51	CoGr 2011 C52 - CoGr 2010 C52
Z	$-,917^b$	$-1,000^b$	$-2,078^b$	$-1,000^b$
Asymp. Sig. (2-tailed)	.359	.317	.038	.317

	CoGr 2011 C53 - CoGr 2010 C53	CoGr 2011 C54 - CoGr 2010 C54	CoGr 2011 C55 - CoGr 2010 C55	CoGr 2011 C56 - CoGr 2010 C56
Z	$-1,890^c$	$-1,841^b$	$-2,144^b$	$-1,725^b$
Asymp. Sig. (2-tailed)	.059	.066	.032	.084

a. Wilcoxon Test

b. Based on negative ranks.

b. Based on positive ranks.

d. The sum of the negative ranks is equal the sum of the positive ranks

Table 6.14: Wilcoxon test statistics for the assessment results for the System-CIP and Point-CIP of the warehouses C29-C56 in the control group

then with the 18 warehouses in the control group that have detailed performance indicators, and finally with the 38 warehouses that do not yet implement performance indicators.

Warehouse Excellence Group versus Control Group (56)

Figure 6.5 reflects the accumulated System-CIP and Point-CIP assessment results of the Warehouse Excellence group in comparison with the control group before the start of the project. Both groups had similar maturity levels. The biggest deviations are in the following areas: Point-CIP with a 0.5 point difference, VSM-Quality with a 0.7 point difference, Sustainable Problem Solving with a 0.6 point difference, and Quality of Problem Solving with a 1.0 point difference. However, the total average score of the control group is 8.6 points and the average total score of the Warehouse Excellence group is 9.2 points. The difference of 0.6 points shows that both groups had similar maturity levels at the beginning of the project because 0.6 points represents 6.5% of the total average score of the Warehouse Excellence group.

However, compared to this development, the gap between the Warehouse Excellence group and the control group is distinctly higher in 2011. Figure 6.6 shows this gap, especially in the Quick Reaction System component which has a difference of 2.0 points and the Improvement Focus, Leadership Involvement, and Regular Communication components which each have a 1.7 point difference. The total score of the Warehouse Excellence group is 32.8 points. The control group has a total score of 12.8 points. This means that a development is recognizable but since it is 20 points lower than the Warehouse Excellence group it is clearly less developed in lean techniques.

Table 6.15 summarizes the comparison of the results of the Warehouse Excellence group and the control group as discussed above. The coefficient of variation has also been listed. Both groups had very similar figures in 2010 but in the year 2011 the Warehouse Ex-

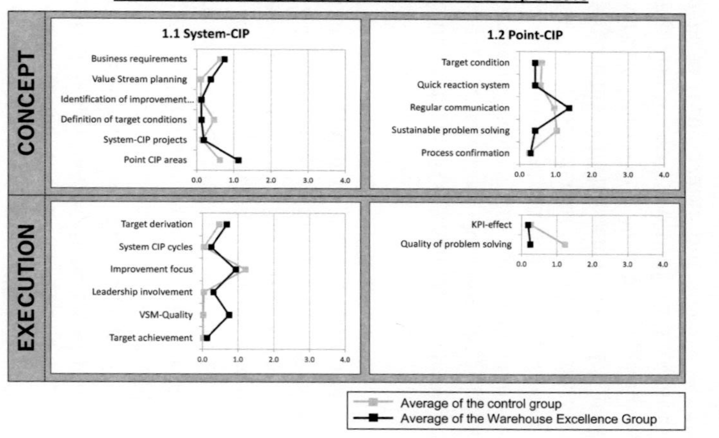

Figure 6.5: BLWA results: Warehouse Excellence group 2010 vs. control group 2010

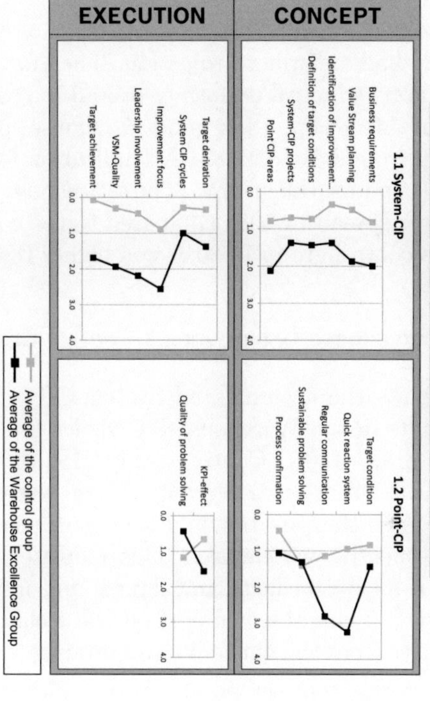

Figure 6.6: BLWA results: Warehouse Excellence group 2011 vs. control group 2011

	WaEx 2010	CoGr 2010	difference 2010	WaEx 2011	CoGr 2011	difference 2011
Average Points in the System- and Piont CIP assessment categories	9,18	8,61	0,57	32,81	12,82	19,99
Variance coefficient in the System- and Point-CIP assessment topics	88,26%	90,85%		28,35%	108,76%	

Table 6.15: Assessment results achieved

cellence group had a distinct projection. The Warehouse Excellence group also improved more uniformly overall in contrast to the control group. In this group, a few good warehouses pulled the average total score up from 8.61 to 12.82. An indication for this is the coefficient of variation for the groups. The Warehouse Excellence group had a narrower spread in 2011 compared to the control group. The spread of the control group in 2011 was higher than in 2010.

Warehouse Excellence Group versus Control Group (18)

Figure 6.7 shows the accumulated System-CIP and Point-CIP assessment results of the Warehouse Excellence group in comparison with control group (18). Control group (18) consists of 18 warehouses from the entire control group. These warehouses have been separated because they are the only ones that measure productivity within the respective warehouse. This indicates that these warehouses have a focuse on facts and figures and might also promote lean techniques. Thus, this comparison will isolate supposedly mature warehouses from the entire control group.

Both groups in this comparison also had similar profiles and maturity levels. However, the gap in the total score achieved between the two groups is broader than in chapter 6.2.2. The total score of the Warehouse Excellence group is 9.2 points and the total score of control group (18) is 10.3 points. The biggest deviations from each other are in the components Improvement Focus with a 1.2 point difference and Quality of Problem solving with 1.1 points difference.

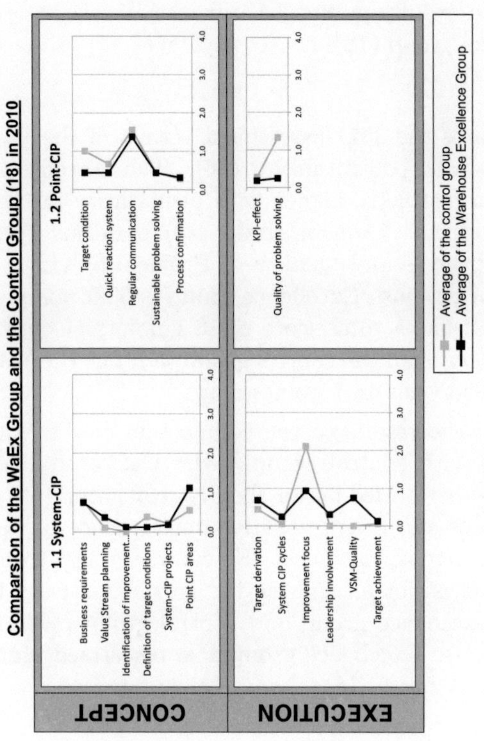

Figure 6.7: BLWA results: Warehouse Excellence group 2010 vs. control group (18) 2010

	WaEx 2010	CoGr (18) 2010	difference 2010	WaEx 2011	CoGr (18) 2011	difference 2011
Average Points in the System- and Piont CIP assessment categories	9,18	10,28	3,9	32,81	18,11	14,7
Variance coefficient in the System- and Point-CIP assessment topics	88,26%	61,06%		28,35%	71,04%	

Table 6.16: BLWA results points: Warehouse Excellence group vs. control group (18)

Figure 6.8 compares the 2011 assessment results of the Warehouse Excellence group with control group (18). The Warehouse Excellence group has a distinctly higher level of maturity in the majority of the components. Control group (18) performs better only in System-CIP Projects and Quality of Problem Solving. The total score of the Warehouse Excellence group is 32.8 points and the control group (18) has a total score of 18.1 points. Control group (18) performs better than the control group (56) but the Warehouse Excellence group still has 14.7 points more.

Table 6.16 shows the results of the comparison of the Warehouse Excellence group with control group (18). The results show that control group (18) performed better than control group (56). Higher average total scores and a less negative development of the coefficient of variation demonstrate this. However, the gap in maturity in this comparison is not as high as the gap in maturity between the Warehouse Excellence group and control group (56). In summary, the Warehouse Excellence group also performed better than the stronger control group (18).

Warehouse Excellence Group versus Control Group (38)

Control group (38) consists of 38 warehouses from the entire control group (56). These warehouses have been separated because they do not measure productivity within the warehouse. We assume that these are the less mature warehouses and we want to complete the

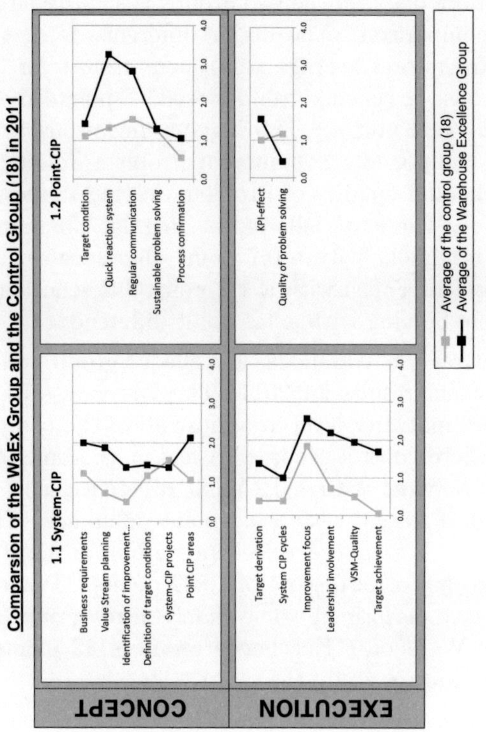

Figure 6.8: BLWA results: Warehouse Excellence group 2011 vs. control group (18) 2011

partial comparison that we began with control group (18). Figure 6.9 shows the accumulated System-CIP and Point-CIP assessment results of the Warehouse Excellence group in comparison with those of control group (38).

Both groups had similar profiles and maturity levels in 2010. The gap in the total score between the two groups is slightly larger than in the previous comparison. The major difference is that control group (38) has lower total average assessment results for the year 2010. The Warehouse Excellence group scored 9.2 points and control group (38) scored 7.82 points. The biggest deviations from each other are in the components Sustainable Problem Solving with a 0.9 point difference and Quality of Problem Solving Process with a 0.8 point difference. These are followed by Regular Communication and Value Stream Quality with a 0.7 point difference each, Point-CIP and Improvement Focus with a 1.3 point difference each, and Quality of Problem Solving with a 1.4 point difference.

Figure 6.10 compares the Warehouse Excellence group and control group (38) assessment results for 2011. The Warehouse Excellence group has a higher maturity level in almost all of the components. The only areas where control group (38) has more points are Sustainable Problem Solving, with a 0.2 point difference, and Quality of Problem Solving Process, with a 0.7 point difference. The total average score of the Warehouse Excellence group is 32.82 points. Control group (38) has a total score of 10.32 points. Control group (38) is found to have performed worse than control group (56) and the gap with the Warehouse Excellence group is 22 points larger than the gap with control group (56).

Table 6.17 summarizes the results of the Warehouse Excellence group in comparison with the control group. It shows that our assumption was correct that the warehouses that do not measure productivity perform with a lower maturity level compared to warehouses with productivity measurements. The results also show that control group (38) did not improve consistently: the coefficient of variation shows this. Both groups had very similar figures in 2010. The Ware-

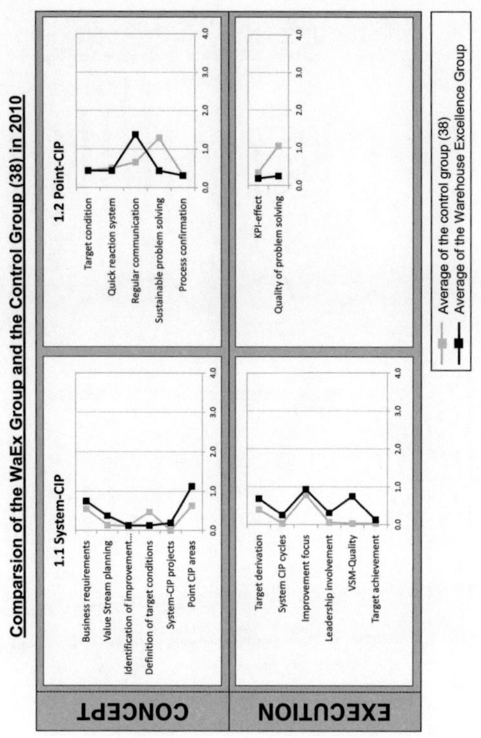

Figure 6.9: BLWA results: Warehouse Excellence group 2010 vs. control group (38) 2010

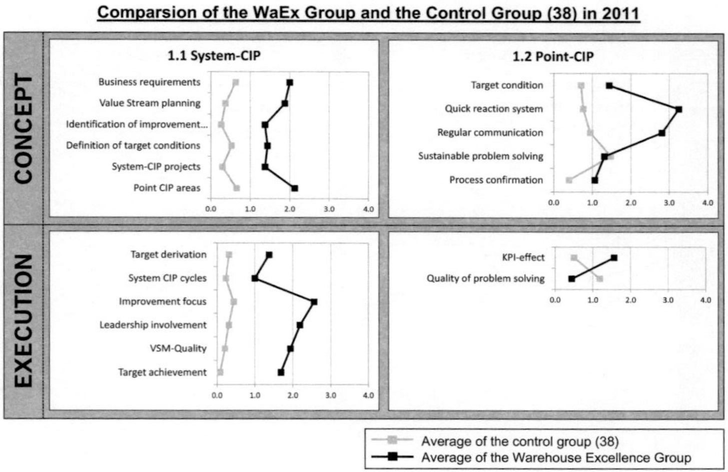

Figure 6.10: BLWA results: Warehouse Excellence group 2011 vs. control group (18) 2011

	WaEx 2010	CoGr (38) 2010	difference 2010	WaEx 2011	CoGr (38) 2011	difference 2011
Average Points in the System- and Piont-CIP assessment topics	9,18	7,82	1,36	32,81	10,32	22,49
Variance coefficient in the System- and Point-CIP assessment topics	88,26%	107,66%		28,35%	134,66%	

Table 6.17: BLWA results points: Warehouse Excellence group vs. control group (38)

house Excellence group had a narrower spread in 2011 compared to the control group. The spread of the control group in 2010 was lower than in 2011.

6.2.3 Intermediate Result: Lean Improvement

Subsection 6.2.1 demonstrated a noticeable improvement in the lean maturity level of the Warehouse Excellence group. That section also showed that the coefficient of variation was lower in 2011 than in 2010. This indicates that the warehouses focused on the lean improvement approach. The better results in the coefficient of variation could be explained with the set milestone goals. Before the project, none of the participating warehouses were mature enough to reach the milestones without an empowerment program. After the empowerment program, all warehouses reached the milestones and fulfilled the set minimum requirements. Since the milestones were set as goals and the warehouses achieved them, a slight tendency towards an alignment of the maturity had taken place.

The notable improvement in the lean maturity levels are clearly shown by the Wilcoxon Signed-Rank test in 6.2.1. The Wilcoxon Signed-Rank test determined that differences between the data sets of 2010 and 2011 exist for 93.75% of the warehouses with a high level of significance. In the control group, the Wilcoxon Signed-Rank test showed that a difference between the samples of the years 2010 and 2011 exists for 21 Warehouses with minimum significance. In conclusion, the percentage of warehouses and significance levels within the control group was lower compared to the Warehouse Excellence group.

The direct comparison of the Warehouse Excellence group with the control group shows that the improvement within the Warehouse Excellence group is higher than the improvement within the control group. The biggest gap in the maturity level is seen in the comparison with the 38 warehouses of the control group which do not measure the productivity, as described in chapter 6.2.2. In turn, the 18 warehouses that measure productivity have the smallest gap, as shown in chapter 6.2.2. Finally, the comparison with the total control group ranks between the two above-mentioned comparisons (see subsection 6.2.2). This leads us to the definition of lean warehousing that is described in chapter 2.4. Part of the philosophy is an analytical approach to driving the continuous improvement process. Analytical approaches are always based on facts and figures. Measuring productivity is a major part in determining the path for improvement. Without the right path, it is difficult to reach a high level of lean maturity. Since we also identify the warehouses that measure productivity as the most mature ones, this indicates that measuring productivity could positively influence improvement. This, in turn, speaks for the quality of the Bosch Logistics Warehouse Assessment that measured the improvement (see chapter 4).

6.3 Analyzing the Impact on Productivity

The development of lean maturity was analyzed in chapter (6.2). The focus of this section is on the development of productivity KPR and KPI. First, the development of the KPR and KPI of the Warehouse Excellence group is analyzed. Then, the KPR development of the Warehouse Excellence group is compared with the KPR development of the control group.

6.3.1 Productivity Development of the Warehouse Excellence Group

Warehouse Excellence Result KPR Productivity Development

From the beginning of the year 2010, warehouses in the Warehouse Excellence group measured the monthly productivity of the entire warehouse operation. Each warehouse reported the monthly average. These monthly average figures were then normalized. This means that the monthly average of January 2010 was set as the index base 100. All further figures were related to that base and represent the development of the original figure. For example, a warehouse had the monthly average productivity of 20 order lines per man hour in January. This would set the index figure at 100. If the figure had a positive development of 10% to 22 order lines per man hour in February, the index would rise to 110. The average of the index developments of the warehouses is shown in figure 6.11.

In 2010, the slope of the trend line was 0.0116. In 2011, the slope rose to the value of 0.0282. The coefficient of the determination of the trend line also rose from 0.2516 to 0.4073. This shows a clear improvement of the KPR in the year 2011. The graph also shows typical seasonal effects on the productivity in summer and winter of those years. These seasonal effects are also an indicator that the graph is reliable.

The Wilcoxon test was used to compare the reported index figures of each warehouse from 2010 with the figures for each warehouse from 2011. The hypotheses are as follows:

H_0:: *The samples n1 and n2 are from the same population*

H_1:: *The samples n1 and n2 are not from the same population*

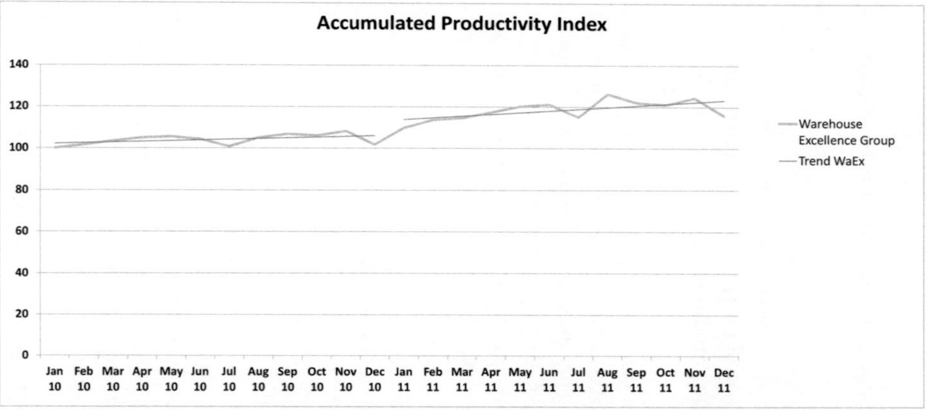

	Jan 10	Feb 10	Mar 10	Apr 10	May 10	Jun 10	Jul 10	Aug 10	Sep 10	Oct 10	Nov 10	Dec 10
Warehouse Excellence Group	100.000	101.532	103.385	104.964	105.465	104.254	100.746	104.914	106.728	106.081	108.163	101.743

	Jan 11	Feb 11	Mar 11	Apr 11	May 11	Jun 11	Jul 11	Aug 11	Sep 11	Oct 11	Nov 11	Dec 11
Warehouse Excellence Group	109.768	113.668	114.616	117.314	120.246	121.140	115.057	126.208	122.157	121.085	124.397	115.949

Figure 6.11: Warehouse Excellence KPR development

		N	Mean Rank	Sum of Ranks
WaEx 2011 - WaEx 2010	Negative Ranks	41[a]	56,00	2296,00
	Positive Ranks	139[b]	100,68	13994,00
	Ties	0[c]		
	Total	180		

a. WaEx 2011 < WaEx 2010

b. WaEx 2011 > WaEx 2010

c. WaEx 2011 = WaEx 2010

Table 6.18: Wilcoxon Rank table for the Warehouse Excellence KPR index in the years 2010 and 2011

The sample n1 are the figures for the year 2010. Sample n2 indicates the sample for the figures in 2011.

The ranking table for the test is shown in table 6.18. It shows that the positive rankings exceed the negative rankings. Finally, the test statistics in 6.18 show with high enough significance that H_0 can be rejected.

Warehouse Excellence Monitoring KPI Productivity Development

During the Warehouse Excellence project, the participating warehouse managers defined projects in specific areas of their warehouses. The goal of these projects was to drive the lean approach, especially the closed loop between the System-CIP and Point-CIP methodologies. Most of the warehouses defined more than one project. However, a minimum of one project was required from the Warehouse Excellence project team. The projects were closely monitored and also followed up on regularly by the project team.

Statistic for Test[a]

	WaEx 2011 - WaEx 2010
Z	-8,355[b]
Asymp. Sig. (2-tailed)	,000

a. Wilcoxon-Test

b. Based on negative ranks.

Table 6.19: Wilcoxon test statistics for the Warehouse Excellence
KPR index in the years 2010 and 2011

Figure 6.12 shows a detailed analysis of each project using one example. The KPI development is highlighted on the left side of the graph. The definition of the KPI is displayed at the top of the figure as is the project name of the warehouse. In this case, the name of the project is W16b. The time line for the project is also shown on this graph. In each example, there is a segment before the beginning of the project. This serves as a basis for comparison. Specific measures in the warehouse that deeply influence productivity are also highlighted. In this example, workforce management was started in November 2011. The average productivity and deviation is listed below for the different segments. The assessment results of the Point-CIP for the years 2010 and 2011 are on the right side. In addition to this, the success factors that the warehouse focused on are also indicated (see section 2.1). Further examples are listed in appendix H.

The 16 projects in the different warehouses had a massive positive influence on the productivity. The improvement was by 26.02% on average. Table 6.20 provides the detailed results of the monitoring KPI development.

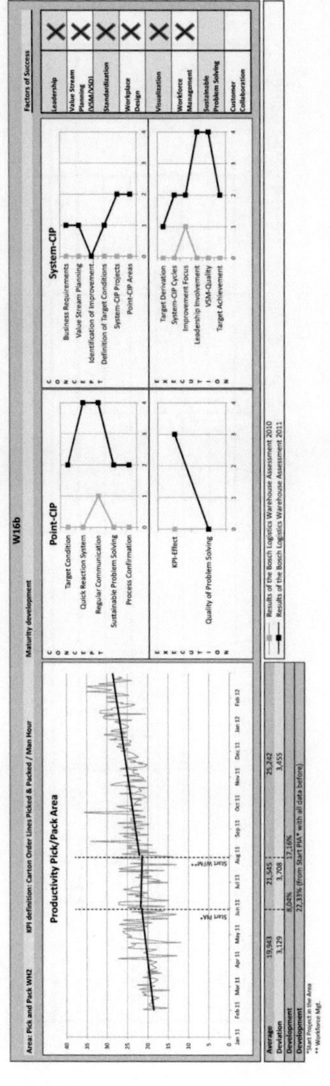

Figure 6.12: Example of monitoring KPI development

Range	X < -25	-25 ≤ X < -15	-15 ≤ X < -5	-5 ≤ X ≤ +5	+5 < X ≤ +15	+15 < X ≤ +25	+25 < X
Amount			2		3	4	7
Result of each warehouse*			-5,61%		5,33%	15,88%	25,39%
			-8,02%		10,52%	21,39%	27,57%
					10,60%	22,33%	33,62%
						23,44%	37,33%
							47,97%
							73,64%
							74,94%

*avarage developement from the defined segment before the project start within the area with
the avarage of after implementation

Table 6.20: Monitoring KPI development overview

6.3.2 Productivity Development of the Warehouse Excellence Group versus Control Group

The KPR and KPI development of the Warehouse Excellence group was shown, analyzed, and interpreted in subsection 6.3.1. This section analyzes the KPR development of the control group. This comparison is shown in figure 6.13. Both of the groups had a similar development in 2010 and the seasonal effects in summer and winter time can be seen. The situation changed in 2011: the control group showed a negative trend. The average index value of 105.26 in 2010 changed to 108.26 in 2011. The trend line slope changed from 0.01463 to -0.02327. In contrast, the Warehouse Excellence group improved its trend from an average of 104.00 in 2010 to an average of 118.47 in 2011. The slope of the trend line increased from 0.01164 to 0.02822.

The non-accumulated index figures of both groups were analyzed using the Kolmogorov-Smirnov Z test to test if improvements are significant or might be coincidences. In the first test, the figures of the Warehouse Excellence group for the year 2010 were compared with the figures of the control group. In the second test, the figures for the year 2011 were compared. The hypotheses for the test are as follows:

H_0: *The samples n1 and n2 are from the same population*

H_1: *The samples n1 and n2 are not from the same population*

The sample n1 indicates the data from 2010. Sample n2 is the data from 2011.

Table 6.21 shows the test frequency. Further results are plotted in table 6.22. In 2010, the significance level is too low to reject H_0. On the other hand, H_0 could be rejected with high significance in

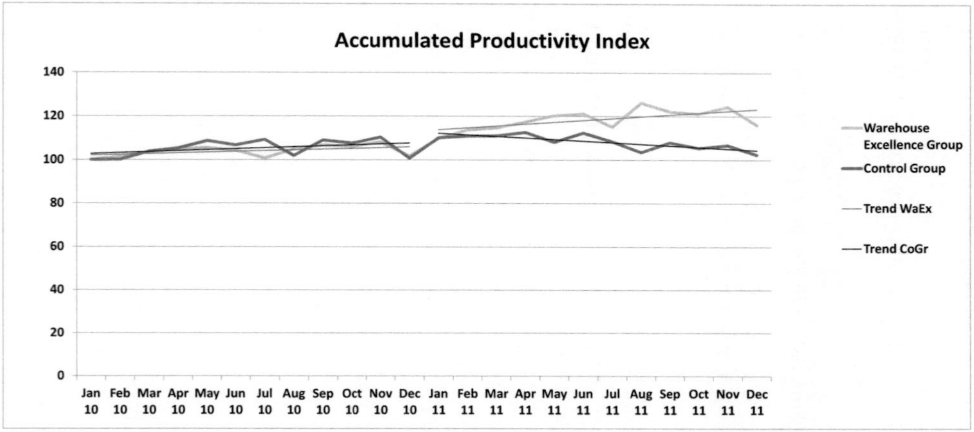

Accumulated Productivity Index

	Jan 10	Feb 10	Mar 10	Apr 10	May 10	Jun 10	Jul 10	Aug 10	Sep 10	Oct 10	Nov 10	Dec 10
Warehouse Excellence Group	100.000	101.532	103.385	104.964	105.465	104.254	100.746	104.914	106.728	106.081	108.163	101.743
Control Group	100.000	100.051	103.784	105.211	108.627	106.667	109.244	101.932	108.948	107.633	110.285	100.712

	Jan 11	Feb 11	Mar 11	Apr 11	May 11	Jun 11	Jul 11	Aug 11	Sep 11	Oct 11	Nov 11	Dec 11
Warehouse Excellence Group	109.768	113.668	114.616	117.314	120.246	121.140	115.057	126.208	122.157	121.085	124.397	115.949
Control Group	110.112	110.879	110.935	112.570	108.072	112.334	108.371	103.474	107.923	105.430	106.657	102.408

Figure 6.13: KPR comparison between the Warehouse Excellence group and control group

Name		N
2010	CoGr	207
	WaEx	180
	Total	387
2011	CoGr	216
	WaEx	192
	Total	408

Table 6.21: Kolmogorov-Smirnov-Test frequency of KPR WaEx vs. CoGr

		2010	2011
most extrem differenz	Absolut	,093	,244
	Positiv	,081	,244
	Negativ	-,093	-,005
Kolmogorov-Smirnov-Z		,910	2,456
Asymp. Sig. (2-tailed)		,379	,000

Table 6.22: Kolmogorov-Smirnov-Test of KPR WaEx vs. CoGr

2011. This means that a large difference in significance can be seen between the data set of the Warehouse Excellence group in 2011 compared to the data set of the control group in 2011.

6.3.3 Intermediate Result: Productivity Improvement

The first analyze of productivity improvement was in subsection 6.3.1 with the Warehouse Excellence KPR. Figure 6.13 shows a higher improvement in year 2011 compared to 2010 and the Kolmogorov-Smirnov-Z test shows, with a high significance, that the data set between 2010 and 2011 is not from the same population. This supports the thesis that an effect could influence KPR development. The analysis in subsection 6.3.1 of the projects carried out during the Warehouse Excellence project tries to explain what the effect could be. The summary shows that a high improvement in the monitoring KPIs has an effect on KPRs.

These positive effects in the Warehouse Excellence group were compared with the control group development in subsection 6.3.2. The graph in figure 6.13 shows a higher productivity development for the Warehouse Excellence group in 2011. The positive trend of the graph is characterized by the Kolmogorov-Smirnov-Z test to show if the result was random or significant. The KPR index graphs from 2010 could be from the same population but in 2011 the graphs show high significance so they are not from the same distribution.

In conclusion, the data and development in 2010 are similar for both groups but are significantly different in 2011 which leads us to the assumption that something happened in the Warehouse Excellence group that did not happen in the control group and it resulted in an improvement of performance. We may suspect that this was the Warehouse Excellence project.

6.4 Review the Hypotheses

We defined the four hypotheses that we wanted to analyze in section 1.2 and they were also explained using a coordinate system in figures 1.1, 1.2, 1.3, and 1.4. The abscissa of the coordinate system shows the lean maturity. The lean maturity improvement was analyzed and an improvement in the Warehouse Excellence group was shown in section 6.2. The ordinate shows the development of the performance indicator. The performance indicator was analyzed in section 6.3. The Warehouse Excellence group showed an improvement in productivity in the result KPR and even a stronger one at the KPI level. We now have the data and the intermediate results we need to make conclusions about the lean impact when discussing the hypotheses and this is done in this section.

Before we start, we will establish the basis for the discussion by showing the relationship between the assessment results and the result KPR of the Warehouse Excellence group in figure 6.14. The

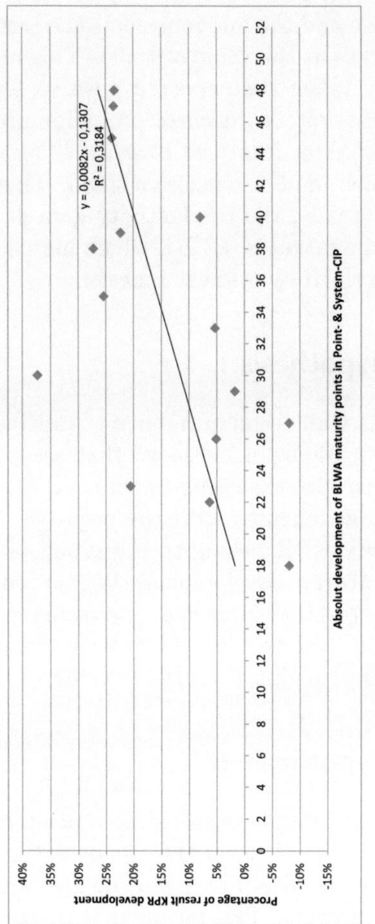

Figure 6.14: Correlation between absolute BLWA System- and Point-CIP development and monitoring KPI developement

113

abscissa is the absolute lean maturity development in the System-CIP and the Point-CIP. The ordinate is the percentage of the result KPR development from the year 2010 to the year 2011. Each point represents one warehouse and the line represents the trend line. The trend line hits the abscissa at the value of 15.94. This indicates that even if some efforts are taken the expected positive lean effect on productivity KPR might not be reached. A minimum higher investment is necessary to gain from the benefits. The slope of the trend line is 0.0082, which implies a positive trend. This shows that if more lean efforts are taken, the productivity gain is also higher. The coefficient of determination is 0.3184, which indicates how well the relation can be described by a linear function.

6.4.1 Review of Hypothesis I

Hypothesis I states that lean techniques have a positive impact on performance indicators. Figure 6.14 shows that most warehouses did have an improvement with the exception of two. The interesting thing is that these two warehouses with the negative development in the result productivity KPR belong to the group of warehouses with the lowest lean maturity development. We can conclude that lean techniques have a positive impact on performance indicators but resources have to be invested into in order to reach a certain lean maturity level before gaining from the benefits.

6.4.2 Review of Hypothesis II

Hypothesis II asserts that more lean has a more positive impact on performance indicators. The slope of the trend line in 6.14 shows that there is a positive relationship between the lean maturity level and the productivity indicators. This means that if you invest more to reach a higher lean maturity you will gain from an associated higher increase in productivity. The figure also shows that if you develop your lean maturity by 30 points, productivity development will increase by a minimum of 5%. To summarize, more lean has a more positive impact on performance indicators.

6.4.3 Review of Hypothesis III

Hypothesis III states that there is a mathematical correlation between the factors lean maturity and performance indicators and a mathematical function can be used to describe this correlation. A relationship between the two factors can be seen in figure 6.14 but a function to describe this correlation could not be found. For example, the coefficient of determination is 0.3184 for a linear regression which is far too low to describe that correlation. In conclusion, a relationship can be identified but a linear mathematical function could not be identified.

6.4.4 Review of Hypothesis IV

Hypothesis IV asserts that lean techniques have a higher positive impact on performance indicators than other approaches. The analysis in section 6.2 showed us that the Warehouse Excellence group improved their lean maturity level significantly: much more than the control group. A large number of the warehouses in the control group did not improve their lean maturity at all so we can assume that the control group warehouses focused on other approaches.

The analysis in subsection 6.3.2 also showed us that the development of the productivity result KPR was similar in both groups in 2010. However, the productivity result KPR development of the Warehouse Excellence group was significantly better than the control group in 2011. This means that something influenced the Warehouse Excellence group in the year 2011. We assume that this relates to the Warehouse Excellence group and that their approach was superior to the other approaches in the control group.

7 Summary & Conclusion

The roots of lean techniques date back 50 years to the production systems of the Japanese automotive production industry. Several in-depth studies have verified the positive impact of lean techniques in production environments. The research methodology of these studies was based on three elements:

- Measurement of the lean maturity
- Measurement of performance indicators
- Comparison of the samples with each other

A high level of evidence about the positive lean impact on performance indicators in the production environment can be proven by comparing the results of these elements with each other. Lean techniques have also found their way into the warehouse environment. Since the warehouse environment is different from the production environment, there is no guarantee that lean techniques have the same impact. Several studies exist on lean maturity, performance indicators or a comparison of different samples with each other but no single study could be found that combines all three elements with each other with the goal of gaining a higher level of evidence about the impact of lean techniques on performance indicators in the warehouse environment. In conclusion, the level of evidence about the positive impact of lean approaches on performance indicators has been higher in the production environment than in the warehouse environment until now.

We carried out a study in an attempt to close this gap in evidence. This study consisted of 16 warehouses in the observation group and 56 warehouses in the control group. The observation group were empowered by an intensive program with precisely defined milestones. By reaching these milestones, it was ensured that the warehouses would implement a structured continuous improvement cycle, which we identified as a key element of the lean philosophy. Bosch coined the terms System-CIP and Point-CIP for their interpretation and definition of the method for a structured continuous improvement cycle process. By providing training, workshops, and coaches; we ensured that all of the warehouses in the observation group reached the set milestones. The control group was not influenced by the empowerment program.

Tools were needed to measure the progress of the lean maturity and performance indicators for each warehouse. By evaluating the existing available tools, we were able to determine that the lean maturity assessments that are customized for the warehouse environment do not meet our requirements. For this reason, a new lean maturity assessment was developed, tested, and implemented. The Bosch Logistic Warehouse Assessment was developed based on a new generation of lean maturity assessments which were in use in the production environment.

The performance indicator development was measured by the KPR/-KPI Tree approach. The KPR/KPI Tree ensures that several measurements can be taken and linked together at the same or different operational level. For example, if a fully developed KRP/KPI Tree was implemented in a warehouse, it would be possible to estimate the influence that the increase in productivity of a picker has on total warehouse productivity. Since this level of development is almost never found within warehouse operations, our study implemented and focused on the result productivity KPR of the total warehouse and the monitoring productivity KPI of specific areas within the warehouse.

After determining what has to be measured and how, we measured and analyzed the generated data with descriptive and inferential statistics. The development of the average assessment score of the observation group was relatively higher than the development of the control group. The two sample non-parametric Wilcoxon hypothesis tests for dependent data were used to test a significantly higher lean maturity development in the observation group compared to the control group.

Before the study began, the development of the total productivity of the observation group in 2010 was very similar to the control group. During the study, the development of the observation group in 2011 was higher than the control group. A significant difference between the groups in 2011 was verified using the two sample non-parametric Kolmogorov-Smirnov hypothesis tests for independent data.

The monitoring productivity KPI of the projects, which is where the strongest impact of the lean activities within the warehouses occurred, also showed a high positive development. A significant functional correlation between the productivity KPR and the lean maturity development could not be verified. Instead, a positive relationship between higher lean maturity and higher productivity gain could be shown. The graph also showed that a certain lean maturity level has to be reached before benefits can be gained from lean.

In conclusion, we have contributed to the evidence that lean techniques have a positive impact on performance indicators. We have also shown that an observation group with a concentrated lean empowerment program performs better than a control group without that focus. A functional correlation between lean techniques and productivity increase could not be shown. This could be because none of the warehouses had a highly developed KPR/KPI Tree. It might be possible to show a correlation with better coverage and a better linking of performance indicators within the observation group and control group.

List of Figures

1.1 Hypothesis I 6
1.2 Hypothesis II 7
1.3 Hypothesis III 8
1.4 Hypothesis IV 9
1.5 Structure of the thesis 11

3.1 Lean Maturity Assessment overview (Fullerton et al.,
     2003)(Hallam, 2003; Pérez and Sánchez, 2000; Paniz-
     zolo, 1998; Shah, 2003; Jordan and Michel, 2001; Doolen
     and Hacker, 2005; Shan, 2008; CMMI Product Team,
     2010; Overboom et al., 2010; Sobanski, 2009; Robert
     Bosch GmbH, 2012) 27
3.2 KPR/KPI Tree example 31

4.1 Bosch Logistics Warehouse Assessment structure . . 37
4.2 Bosch Logistic Warehouse Assessment Topics Part 1 41
4.3 Bosch Logistic Warehouse Assessment Topics Part 2 42
4.4 BLWA link between execution and concept 44
4.5 BLWA example for linked criteria 45

5.1 Warehouse Excellence empowerment structure . . . 51
5.2 Warehouse Excellence project milestones 52
5.3 Proof structure of the thesis 57

6.1 BLWA results: Warehouse Excellence group 2010 vs.
     Warehouse Excellence group 2011 (part 1) 66

6.2   BLWA results: Warehouse Excellence group 2010 vs. Warehouse Excellence group 2011 (part 2)   67

6.3   BLWA results: Warehouse Excellence group 2010 vs. Warehouse Excellence group 2011 (part 3)   68

6.4   BLWA results: Warehouse Excellence group 2010 vs. Warehouse Excellence group 2011 (part 4)   69

6.5   BLWA results: Warehouse Excellence group 2010 vs. control group 2010   92

6.6   BLWA results: Warehouse Excellence group 2011 vs. control group 2011   93

6.7   BLWA results: Warehouse Excellence group 2010 vs. control group (18) 2010   95

6.8   BLWA results: Warehouse Excellence group 2011 vs. control group (18) 2011   97

6.9   BLWA results: Warehouse Excellence group 2010 vs. control group (38) 2010   99

6.10   BLWA results: Warehouse Excellence group 2011 vs. control group (18) 2011   100

6.11   Warehouse Excellence KPR development   104

6.12   Example of monitoring KPI development   107

6.13   KPR comparison between the Warehouse Excellence group and control group   110

6.14   Correlation between absolute BLWA System- and Point-CIP development and monitoring KPI developement   113

C.1   1.1 System-CIP Concept (Furmans and Wlcek, 2012; Rother and Shook, 2008; Sobanski, 2009; Dehdari et al., 2011)   142

C.2   1.1 System-CIP Execution (Furmans and Wlcek, 2012; Sobanski, 2009; Rother and Shook, 2008)   143

C.3   1.2 Point-CIP Concept (Dehdari et al., 2011; Sobanski, 2009)   144

C.4   1.2 Point-CIP Execution (Sobanski, 2009; Dehdari et al., 2011)   145

C.5   2.1 Failure Prevention System (Hoyle, 2006;
        Vahrenkamp, 2010)   146
C.6   2.2 Employee Involvement (Jackson and Jones, 1996;
        Sobanski, 2009)   147
C.7   2.3 Standardized Work (Sobanski, 2009; Graupp and
        Wrona, 2006)   148
C.8   3.1 Overhead (Sobanski, 2009)   149
C.9   3.2 Outgoing Goods (Dehdari and Schwab, 2012; Soban-
        ski, 2009; Furmans and Wlcek, 2012)   150
C.10 3.3 Packaging   151
C.11 3.4 Picking   152
C.12 3.5 Storage   153
C.13 3.6 Incoming Goods Concept (Dehdari and Schwab,
        2012; Furmans and Wlcek, 2012)   154
C.14 3.6 Incoming Goods Excecution (Furmans and Wlcek,
        2012; Sobanski, 2009)   155

H.1   Project development sheet Warehouse 1   176
H.2   Project development sheet Warehouse 2   177
H.3   Project development sheet Warehouse 4   178
H.4   Project development sheet Warehouse 5   179
H.5   Project development sheet Warehouse 6   180
H.6   Project development sheet Warehouse 7   181
H.7   Project development sheet Warehouse 8   182
H.8   Project development sheet Warehouse 9   183
H.9   Project development sheet Warehouse 10   184
H.10 Project development sheet Warehouse 11   185
H.11 Project development sheet Warehouse 12   186
H.12 Project development sheet Warehouse 13   187
H.13 Project development sheet Warehouse 14   188
H.14 Project development sheet Warehouse 15   189
H.15 Project development sheet Warehouse 16   190
H.16 Project development sheet Warehouse 16b   191

List of Tables

3.1 Lean Production efficiency studies (Author's illustration based on Doolen and Hacker, 2005) (Inman and Mehra, 1993; Biggart, 1997; Claycomb et al., 1999; Fullerton and McWatters, 2001; Germain et al., 1996; Kinney and Wempe, 2002; Fullerton et al., 2003; Matsui, 2007; Jayaram et al., 2008; Fullerton and Wempe, 2009; Yang et al., 2011; Hofer et al., 2012; ?)   23

6.1 BLWA results: Warehouse Excellence group 2010 vs. Warehouse Excellence group 2011 average scores (part 1)   70
6.2 BLWA results: Warehouse Excellence group 2010 vs. Warehouse Excellence group 2011 average scores (part 2)   71
6.3 BLWA results: total points   72
6.4 Shapiro-Wilk test for the total assessment results of the Warehouse Excellence group   75
6.5 Shapiro-Wilk test for the System-CIP and Point-CIP assessment results of the warehouses in the Warehouse Excellence group   77
6.6 Shapirot-Wilk test for the assessment results of the control group   78
6.7 Wilcoxon Rank table for the assessment results of the warehouses in the Warehouse Excellence group . . .   81
6.8 Wilcoxon test statistics for the assessment results of the warehouses in the Warehouse Excellence group .   82

6.9 Wilcoxon Rank table for the assessment results for the System-CIP and Point-CIP of the warehouses in the Warehouse Excellence group 83

6.10 Wilcoxon test statistics for the assessment results for the System-CIP and Point-CIP of the warehouses in the Warehouse Excellence group 85

6.11 Wilcoxon Rank table for the assessment results for the System-CIP and Point-CIP of the warehouses C1-C30 in the control group 86

6.12 Wilcoxon Rank table for the assessment results for the System-CIP and Point-CIP of the warehouses C31-C56 in the control group 87

6.13 Wilcoxon test statistics for the assessment results for the System-CIP and Point-CIP of the warehouses C1-C28 in the control group 89

6.14 Wilcoxon test statistics for the assessment results for the System-CIP and Point-CIP of the warehouses C29-C56 in the control group 90

6.15 Assessment results achieved 94

6.16 BLWA results points: Warehouse Excellence group vs. control group (18) 96

6.17 BLWA results points: Warehouse Excellence group vs. control group (38) 101

6.18 Wilcoxon Rank table for the Warehouse Excellence KPR index in the years 2010 and 2011 105

6.19 Wilcoxon test statistics for the Warehouse Excellence KPR index in the years 2010 and 2011 106

6.20 Monitoring KPI development overview 108

6.21 Kolmogorov-Smirnov-Test frequency of KPR WaEx vs. CoGr 111

6.22 Kolmogorov-Smirnov-Test of KPR WaEx vs. CoGr 111

A.1 Classification of warehouses in the Warehouse Excellence group 138

B.1 Classification of warehouses in the control group . . 140

D.1 BLWA results per warehouse: Warehouse Excellence
group 2010 158
D.2 BLWA results per warehouse: Warehouse Excellence
group 2011 159

E.1 Warehouse Excellence Group KPR development 2010 162
E.2 Warehouse Excellence KPR development 2011 163

F.1 Assessment results per warehouse: control group 2010
C1-C28 166
F.2 Assessment results per warehouse: control group 2010
C29-C56 167
F.3 Assessment results per warehouse: control group 2011
C1-C28 168
F.4 Assessment results per warehouse: control group 2011
C1-C28 169

G.1 Control group KPR development 2010 172
G.2 Control group KPR development 2011 173

Bibliography

Anderson, T. W. and D. A. Darling (1954). A Test of Goodness of Fit. *Journal of the American Statistical Association 49*(268), p. 765–769.

Arnold, D., H. Isermann, A. Kuhn, H. Tempelmeier and K. Furmans (2008). *Logistik-Handbuch*. Berlin and Heidelberg: Springer Berlin Heidelberg.

Augustin, H. (2009). *Lean warehousing: Mit Six Sigma und Lean Management Lagerprozesse erfolgreich gestalten ; eine Trendstudie*. München: Huss.

Backhaus, K., B. Erichson, W. Plinke and R. Weiber (2003). *Multivariate Analysemethoden: Eine anwendungsorientierte Einführung* (10 ed.). Berlin: Springer.

Bartholdi, J. J. and S. T. Hackman (2011). Warehouse & Distribution Science.

Bell, S. and M. A. Orzen (2011). *Lean IT: enabling and sustaining your lean transformation*. New York: Productivity Press.

Bidgoli, H. (eds.) (2004). *Managing the Flow of Materials Across the Supply Chain // The Internet encyclopedia*. Hoboken and NJ and USA: John Wiley & Sons, Inc and John Wiley & Sons.

Biggart, T. (1997). *The Effects of Just-in-Time Inventory System Adoption on Firm Performance*. Ph.D. thesis.

Bol, G. (2003). *Induktive Statistik: Lehr- und Arbeitsbuch* (3 ed.). München and Wien: Oldenbourg.

Bortz, J. and R. Weber (2005). *Statistik für Human- und Sozialwissenschaftler: Mit 242 Tabellen* (6 ed.). Springer-Lehrbuch. Heidelberg: Springer Medizin.

Bösenberg, D. and H. Metzen (1993). *Lean Management: Vorsprung durch schlanke Konzepte* (2 ed.). Landsberg/Lech: Verl. Moderne Industrie.

Brosius, F. (2011). *SPSS 19* (1 ed.). Heidelberg and München and Landsberg and Frechen and Hamburg: mitp-Verl.

Bruin, T. d., R. Freeze, U. Kaulkarni and M. Rosemann (eds.) (2005). *Understanding the Main Phases of Developing a Maturity Assessment Model.* Australasian Conference on Information Systems. Sydney: Campbell, B. and Underwood, J. and Bunker, D.

Cirrone, G., S. Donadio, S. Guatelli, A. Mantero, B. Mascialino, S. Parlati, M. Pia, A. Pfeiffer, A. Ribon and P. Viarengo (2004). A goodness-of-fit statistical toolkit. *IEEE Transactions on Nuclear Science 51*(5), p. 2056–2063.

Clauß, G., F.-R. Finze and L. Partzsch (1994). *Statistik für Soziologen, Pädagogen, Psychologen und Mediziner.* Thun [u.a.]: Deutsch.

Claycomb, C., C. Dröge and R. Germain (1999). The Effect of Just-in-Time with Customers on Organizational Design and Performance. *The International Journal of Logistics Management 10*(1), p. 37–58.

Claycomb, C., R. Germain and C. Dröge (1999). Total system JIT outcomes: inventory, organization and financial effects. *International Journal of Physical Distribution & Logistics Management 29*(10), p. 612–630.

CMMI Product Team (2010). CMMI® for Services, Version 1.3: CMMI-SVC, V1.3.

Cusumano, M. A. (1994). The Limits of Lean. *Sloan Management Review Summer*, p. 27–32.

Dehdari, P. and M. Schwab (2012). Produktionsumfeld vs. Lagerumfeld. In: K. Furmans and H. Wlcek (eds.), *Arbeitskreisbericht*, Volume 1, p. 16–20. DVV Media Group.

Dehdari, P., M. Schwab, K. Furmans and H. Wlcek (2011). Können Läger schlank sein? Von Lean Production über Lean Management zum Lean Warehousing. In: T. Wimmer (eds.), *Flexibel - sicher - nachhaltig.* Hamburg: DVV Media Group.

Dehdari, P., M. Schwab, K. Furmans and H. Wlcek (2012). Lean Assessment für das Lagerumfeld: - „Entschuldigung, wo geht es hier zum Leuchtturm?". *Productivity Management 5*, p. 23–27.

Doolen, T. L. and M. E. Hacker (2005). A review of lean assessment in organizations: An exploratory study of lean practices by electronics manufacturers. *Journal of Manufacturing Systems 24*(1), p. 55–67.

Duller, C. (eds.) (2008). *Einführung in die nichtparametrische Statistik mit SAS und R: Ein anwendungsorientiertes Lehr- und Arbeitsbuch.* Heidelberg: Physica-Verlag Heidelberg.

Dykes, n. (1999). Debunking Popper: A Critique of Karl Popper's Critical Rationalism. *Reason Papers* (24), p. 5–26.

EN 14943 (2005). Transport services - Logistics - Glossary of terms.

Flinchbaugh, J. (2009). Untangling spaghetti. *Assembly 52*(2), p. 75.

Fullerton, R. R. and C. S. McWatters (2001). The production performance benefits from JIT implementation. *Journal of Operations Management 19*(1), p. 81–96.

Fullerton, R. R., C. S. McWatters and C. Fawson (2003). An examination of the relationships between JIT and financial performance. *Journal of Operations Management 21*(4), p. 383–404.

Fullerton, R. R. and W. F. Wempe (2009). Lean manufacturing, non-financial performance measures, and financial performance. *International Journal of Operations & Production Management 29*(3), p. 214–240.

Furmans, K. and H. Wlcek (eds.) (2012). *Arbeitskreisbericht: Lean Management in Lägern.* DVV Media Group.

Genschel, U. and C. Becker (2005). *Schließende Statistik: Grundlegende Methoden.* Berlin/Heidelberg: Springer-Verlag.

Germain, R., C. Droge and N. Spears (1996). The implications of just-in-time for logistics organization management and performance. *Journal of Business Logistics 17 (2)*, p. 19–34.

Gibbons, J. D. (2003). *Nonparametric statistical inference.* New York: Marcel Dekker.

Graupp, P. and R. J. Wrona (2006). *The TWI workbook: Essential skills of supervisors.* New York: Productivity Press.

Gudehus, T. and H. Kotzab (2012). *Comprehensive logistics* (2 ed.). Heidelberg and and New York: Springer-Verlag Berlin Heidelberg.

Hallam, C. R. A. (2003). *Lean enterprise self-assessment as a leading indicator for accelerating transformation in the aerospace industry.* Ph.D. thesis, Massachusetts Institute of Technology (MIT), Massachusetts.

Hofer, C., C. Eroglu and A. Rossiter Hofer (2012). The effect of lean production on financial performance: The mediating role of inventory leanness. *International Journal of Production Economics 138*(2), p. 242–253.

Holweg, M. (2007). The genealogy of lean production. *Journal of Operations Management 25*(2), p. 420–437.

Hoyle, D. (2006). *ISO 9000 quality systems handbook* (5 ed.). Oxford U.K. and and Burlington MA: Butterworth-Heinemann.

Huber, C. (2011). *Throughput Analysis of Manual Order Picking Systems with Congestion Consideration.* Ph.D. thesis, Karlsruhe.

Inman, R. A. and S. Mehra (1993). Financial Justification of JIT Implementation. *International Journal of Operations & Production Management 13*(4), p. 32–39.

Jackson, T. L. and K. R. Jones (1996). *Implementing a lean management system.* Portland Or: Productivity Press.

Janssen, J. (2005). *Statistische Datenanalyse mit SPSS für Windows: Eine anwendungsorientierte Einführung in das Basissystem und das Modul Exakte Tests; mit 163 Tabellen* (5 ed.). Berlin and Heidelberg: Springer.

Jayaram, J., S. Vickery and C. Droge (2008). Relationship building, lean strategy and firm performance: an exploratory study in the automotive supplier industry. *International Journal of Production Research 46*(20), p. 5633–5649.

Jordan, J. A. and F. J. Michel (2001). *The lean company: Making the right choices.* Dearborn and MI: Society of Manufacturing Engineers.

Kinney, M. R. and W. F. Wempe (2002). Further Evidence on the Extent and Origins of JIT's Profitability Effects. *The Accounting Review 77*(1), p. 203–225.

Lass, R. (1984). *Phonology: An introduction to basic concepts.* Cambridge and and New York: Cambridge University Press.

Lebra, T. S. and W. P. Lebra (1986). *Japanese culture and behavior: Selected readings* (Rev. ed. ed.). Honolulu: University of Hawaii Press.

Lehmann, E. L. and J. P. Romano (2005). *Testing statistical hypotheses* (3 ed.). Springer texts in statistics. New York: Springer.

Liker, J. K. (2004). *The Toyota way: 14 management principles from the world's greatest manufacturer.* New York: McGraw-Hill.

Mahfouz, A. (2011). *An Integrated Framework to Assess 'Leanness' Performance in Distribution Centres.* Ph.D. thesis, Dublin Institute of Technology, Dublin.

Matsui, Y. (2007). An empirical analysis of just-in-time production in Japanese manufacturing companies. *International Journal of Production Economics 108*(1-2), p. 153–164.

Oeltjenbruns, H. (2000). *Organisation der Produktion nach dem Vorbild Toyotas: Analyse Vorteile und detaillierte Voraussetzungen sowie die Vorgehensweise zur erfolgreichen Einführung am Beispiel eines globalen Automobilkonzerns: Zugl.: Clausthal, Techn. Univ., Diss., 2000.* Innovationen der Fabrikplanung und -organisation. Aachen: Shaker.

Ohno, T. (1988). *Toyota production system: Beyond large-scale production*. Cambridge and Massachusetts: Productivity Press.

Olhager, J. (2012). The Role of Decoupling Points in Value Chain Management. In: H. Jodlbauer, J. Olhager, and R. J. Schonberger (eds.), *The Role of Decoupling Points in Value Chain Management*, p. 37–47. Heidelberg: Physica-Verlag HD.

Overboom, M. A., de Haan J. A. C. and Naus A. J. A. M. (eds.) (2010). *Measuring the degree of leanness in logistics service providers: Development of a measurement tool*, Porto.

Panizzolo, R. (1998). Applying the lessons learned from 27 lean manufacturers. *International Journal of Production Economics 55*(3), p. 223–240.

Pérez, M. P. and A. M. Sánchez (2000). Lean production and supplier relations: a survey of practices in the Aragonese automotive industry. *Technovation 20*(12), p. 665–676.

Pfeiffer, W. and E. Weiß (1994). *Lean Management. Grundlagen der Führung und Organisation lernender Unternehmen. 2., überarb. und erw. Aufl.* Berlin: Schmidt.

Razali, N. M. and Y. B. Wah (2011). Power comparisions of Shapiro-Wilk, Kolmogorov-Smirnov, Lilliefors and Anderson-Darling tests. *Journal of Statistical Modeling and Analytics 2*(1), p. 21–33.

Reuter, C. (2009). *Logistikrelevante Lösungen auf der Basis von Lean-Management bei kleinen Losgrössen und hoher Variantenvielfalt.* Ph.D. thesis, Universität Stuttgart, Heimsheim.

Robert Bosch GmbH (2012). Bosch Production System Assessment V. 3.1.

Rother, M. and S. Kinkel (2009). *Die Kata des Weltmarktführers: Toyotas Erfolgsmethoden* (1 ed.). Frankfurt: Campus Verl.

Rother, M. and J. Shook (2008). *Learning to see: Value-stream mapping to create value and eliminate muda* (1.3 ed.). A lean tool kit method and workbook. Cambridge and Massachusetts: Lean Enterprise Inst.

Sá, J. P. M. d. (eds.) (2008). *Applied Statistics Using SPSS, STATIS-TICA, MATLAB and R.* Berlin and Heidelberg: Springer-Verlag.

Scheer, A.-W. (eds.) (2005). *Corporate Performance Management: ARIS in der Praxis; mit 5 Tab.* Berlin and Heidelberg: Springer.

Shah, R. (2003). Lean manufacturing: context, practice bundles, and performance. *Journal of Operations Management 21*(2), p. 129–149.

Shan, R. S. (2008). *The role of assessment in a lean transformation.* Ph.D. thesis, Massachusetts Institute of Technology (MIT), Massachusetts.

Sobanski, E. B. (2009). *Assessing Lean Warehousing: Development and validation of a lean assessment tool.* Ph.D. thesis, Oklahoma State University, Oklahoma.

Spee, D. and J. Beuth (2012). *Lagerprozesse effizient gestalten: Lean Warehousing in der Praxis erfolgreich anwenden.* München: Huss-Verl.

ten Hompel, M., T. Schmidt and L. Nagel (2007). *Materialflusssysteme: Förder- und Lagertechnik* (3., vollst. neu bearb. Aufl ed.). Berlin: Springer.

Toutenburg, H., C. Heumann, M. Schomaker and M. Wißmann (eds.) (2009). *Arbeitsbuch zur deskriptiven und induktiven Statistik* (2 ed.). Springer-Lehrbuch. Berlin and Heidelberg: Springer Berlin Heidelberg.

Vahrenkamp, R. (2010). *Von Taylor zu Toyota: Rationalisierungsdebatten im 20. Jahrhundert* (1 ed.). Lohmar and Köln: Eul.

VDI 3629 (2005). Basic organisational functions in warehousing.

Wisser, J. (2009). *Der Prozess Lagern und Kommissionieren im Rahmen des Distribution Center Reference Model (DCRM).* Wissenschaftliche Berichte des Institutes für Fördertechnik und Logistiksysteme der Universität Karlsruhe (TH). Karlsruhe: Universitätsverlag.

Womack, J. P. (2007). *The machine that changed the world: The story of lean production - Toyotas secret weapon in the global car wars that is revolutionizing world industry* (1 ed.). London: Free Press.

Womack, J. P. and D. T. Jones (2004). *Lean thinking: Ballast abwerfen, Unternehmensgewinne steigern* ([Erw. und aktualisierte Neuausg.] ed.). Management. Frankfurt am Main: Campus-Verl.

Yang, M. G., P. Hong and S. B. Modi (2011). Impact of lean manufacturing and environmental management on business performance: An empirical study of manufacturing firms. *International Journal of Production Economics 129*(2), p. 251–261.

Zidel, T. (2006). *A lean guide to transforming healthcare: How to implement lean principles in hospitals, medical offices, clinics, and other healthcare organizations.* Milwaukee and Wis: ASQ Quality Press.

A Appendix – Warehouse Excellence Group Data Sheet

WH* Code	Region**	Business Unit	Service Provider	WH Type***	Clients beside Bosch	# Staff	Storage Capacity		Inbound Vol./Day		Outbound Vol./Day	
							# Pallets	# Bins	# Trucks	# Pallets	# Trucks	# Pallets
W1	EMEA	UBK	LSP	D	Yes	60	22.000	17.000	15	350	20	250
W2	EMEA	UBK	Bosch	P	No	160	2.520	21.600	7	110	40	350
W3	EMEA	UBI	Bosch	D	No	70	2.500	46.000	10	50	12	150
W4	EMEA	UBG	LSP	D	Yes	75	40.000	16.000	70	1.000	40	800
W5	EMEA	UBI	LSP	P	Yes	80	40.000	30.000	40	1.000	40	1.000
W6	EMEA	UBK	Bosch	P	No	330	4.500	10.500	70	400	50	300
W7	EMEA	UBG	LSP	D	Yes	75	15.000	20.000	40	N/A	40	N/A
W8	EMEA	UBK	LSP	P	Yes	35	12.000	2.500	25	400	40	600
W9	EMEA	UBK	Bosch	P	No	220	9.500	15.000	40	800	30	600
W10	EMEA	UBK/UBG	LSP	D	No	120	12.900	30.000	4	210	10	220
W11	EMEA	UBK	LSP	P	Yes	100	1.000	10.000	N/A	200	N/A	350
W12	EMEA	UBK	LSP	P	Yes	45	15.000	N/A	35	N/A	70	N/A
W13	EMEA	UBG	LSP	P	Yes	150	25.000	13.800	75	700	45	700
W14	EMEA	UBK	Bosch	P	No	36	3.300	6.000	20	450	15	350
W15	EMEA	UBK	LSP	P	No	16	2.500	6.000	3	150	6	180
W16	EMEA	UBK/UBG	Bosch	D	No	125	14.000	60.000	4-6	400	4-6	N/A

* WH = Warehouse

** EMEA = Europe/Middle-East/Africa

*** D = Distribution

P = Production

Table A.1: Classification of warehouses in the Warehouse Excellence group

B Appendix – Control Group Data Sheet

Warehouse Code	Region	Business Unit	Service Provider	Warehouse Type	Storage Capacity # Pallets
C1	Asia-Pacific	UBK	LSP	Production	6.000
C2	Asia-Pacific	UBK	LSP	Production	9.000
C3	Asia-Pacific	UBI	LSP	Distribution	4.891
C4	Asia-Pacific	UBK	LSP	Distribution	2.922
C5	Asia-Pacific	UBK	LSP	Distribution	12.640
C6	Asia-Pacific	UBI	LSP	Distribution	12.000
C7	Asia-Pacific	UBG	Bosch	Distribution	4.500
C8	Asia-Pacific	UBK	LSP	Production	5.000
C9	Asia-Pacific	UBK	Bosch	Distribution	1.353
C10	Asia-Pacific	UBK	Bosch	Distribution	5.672
C11	Asia-Pacific	UBK	Bosch	Distribution	911
C12	Asia-Pacific	UBI	Bosch	Production	3.709
C13	Asia-Pacific	UBK	Bosch	Distribution	430
C14	Asia-Pacific	UBK	Bosch	Distribution	950
C15	Asia-Pacific	UBK	Bosch	Distribution	595
C16	Asia-Pacific	UBK	Bosch	Distribution	207
C17	Asia-Pacific	UBK	Bosch	Distribution	383
C18	Asia-Pacific	UBK	Bosch	Distribution	947
C19	Asia-Pacific	UBK	Bosch	Distribution	1.166
C20	Asia-Pacific	UBK	Bosch	Distribution	1.984
C21	Asia-Pacific	UBK	Bosch	Distribution	1.118
C22	Asia-Pacific	UBK	Bosch	Distribution	896
C23	Asia-Pacific	UBK	Bosch	Distribution	633
C24	Asia-Pacific	UBK	Bosch	Distribution	721
C25	Asia-Pacific	UBK	Bosch	Distribution	4.058
C26	Asia-Pacific	UBK	Bosch	Distribution	1.732
C27	Europe/Middle East/Africa	UBK	LSP	Distribution	5.255
C28	Europe/Middle East/Africa	UBG	LSP	Distribution	3.900
C29	Europe/Middle East/Africa	UBG	LSP	Distribution	4.800
C30	Europe/Middle East/Africa	UBK	LSP	Production	3.000
C31	Europe/Middle East/Africa	UBK	LSP	Distribution	800
C32	Europe/Middle East/Africa	UBK	LSP	Distribution	6.652
C33	Europe/Middle East/Africa	UBK	LSP	Distribution	1.018
C34	Europe/Middle East/Africa	UBG	LSP	Distribution	12.800
C35	Europe/Middle East/Africa	UBK	LSP	Distribution	N/A
C36	Europe/Middle East/Africa	UBK	LSP	Production	5.800
C37	Europe/Middle East/Africa	UBG	LSP	Distribution	14.100
C38	Europe/Middle East/Africa	UBI	Bosch	Production	10.433
C39	Europe/Middle East/Africa	UBG	LSP	Production	11.400
C40	Europe/Middle East/Africa	UBK	Bosch	Production	7.600
C41	Europe/Middle East/Africa	UBI	LSP	Distribution	9.300
C42	Europe/Middle East/Africa	UBG	LSP	Production	12.000
C43	Europe/Middle East/Africa	UBG	LSP	Production	18.000
C44	Europe/Middle East/Africa	UBK	LSP	Distribution	9.230
C45	Europe/Middle East/Africa	UBK	LSP	Distribution	3.000
C46	Europe/Middle East/Africa	UBG	Bosch	Production	7.550
C47	Europe/Middle East/Africa	UBG	Bosch	Distribution	5.250
C48	Europe/Middle East/Africa	UBG	LSP	Distribution	16.000
C49	Europe/Middle East/Africa	UBK	LSP	Distribution	1.180
C50	Europe/Middle East/Africa	UBK	LSP	Production	850
C51	Latin America	UBK	Bosch	Production	3.120
C52	Latin America	UBK	LSP	Distribution	600
C53	North America	UBI	Bosch	Distribution	3.344
C54	North America	UBK	Bosch	Production	10.000
C55	North America	UBK	LSP	Production	8.500
C56	North America	UBG	Bosch	Production	3.389

Table B.1: Classification of warehouses in the control group

C Appendix - Assessment Questionaire

The development of the Bosch Logistic Warehouse Assessment (BLWA) is described in chapter4. We remember that the BLWA is based on the Bosch Production System Assessment V. 3.1. Some parts of the Bosch Production System Assessment V. 3.1. were used as is, some parts derived, and some parts developed new for the BLWA. However, the Bosch Production System Assessment V. 3.1 is the intellectual property of Bosch and classified as strictly confidential. This means that parts that are used as is or derived cannot be published and only the parts that are totally new could be published. Nevertheless, we looked for literature sources that explain the main purpose of the parts that cannot be published. These parts were rated from bad to very good.

| Warehouse Analysis 1.0 | | | | Level | | | | | Points | Comments |
No.	Topics	Description	0	1 Standards	2 Standards	3 Standards	4 Standards			
1.1	System CIP / CONCEPT	**Business requirements** To establish a vision should be an integral part of the lean practise in a warehouse. (Sobanski 2009, p.210) To reach this vision, KPI-Trees can be used to derive goals from the business (customer/market) for the warehouse. (Furmans, 2012, p. 79)		bad	ok	good	very good		0	
		Value stream planning Sobanski (2009, p.210) explains the importance of value stream mapping. A higher maturity can be reached by having value stream mapping and value stream design for all processes within the warehouses. The value stream maps should include the information and material flows and key performance indicators. (Rother, 2008, part II). A yearly update of the maps are necessary (Rother, 2008, part V). In a warehouse with higher maturity, the role of the value stream manager should be implemented and the value stream design should lead to a pull system. (Rother, 2008, part I & III)		bad	ok	good	very good		0	
		Identification of improvement activities Rother (2008, part V) mentions that the improvement activities should lead from value stream mapping to value stream design. These activities need to have measurable goals. Also, a derivation from the business requirement could help identify successful improvement activities. (Furmans, 2012, p.80)		bad	ok	good	very good		0	
		Definition of target conditions for Point-CIP Dehdari et al. (2011) mention that a target condition with a high maturity consist of a standard, a key performance indicator, and stability criteria.		bad	ok	good	very good		0	
		System-CIP projects Rother (2008, part V) describes the following requirements for projects. First, it has to be exactly described what you plan to do, when, and how (step-by-step). Then, measurable goals are needed. Finally, clear checkpoints with real deadlines and named reviewers are required. He also highlight also the importance of the value stream manager within this context.		bad	ok	good	very good		0	
				Point-CIP areas: - One Point-CIP for whole warehouse	Point CIP areas: - Point CIP done for minimum two warehouse sections	Point CIP areas: - Point CIP done separately for every warehouse section (IG, Storage, ...)	Point CIP areas: - Point CIP done separately for every warehouse section, circa 10 operators per team			
		Average							0.00	

Figure C.1: 1.1 System-CIP Concept (Furmans and Wlcek, 2012; Rother and Shook, 2008; Sobanski, 2009; Dehdari et al., 2011)

			Description	KPI	KPI	KPI	KPI		
1.1	System CIP	EXECUTION	**Target derivation** The management has to set the goals. (Furmans, 2012, p.82)	bad	ok	good	very good	0	
			System-CIP cycles A higher maturity can be reached by the number of formal annual System-CIP cycles (Kaizen) events conducted at the facility. Sobanski (2008, p. 234) speaks from 1 to 10.	bad	ok	good	very good	0	
			Improvement focus Furmans (2012, p. 81) says that the improvement should be in the key performance indicators of quality, delivery performance, and costs.	bad	ok	good	very good	0	
			Leadership involvement Rother (2008, part V) mentions that the role of the manager is to know the value stream and drive the improvement work forward.	bad	ok	good	very good	0	
			Value stream quality It is important to use standardized icons to have the same understanding of the value stream. More information like process KPIs leads to a more mature value stream quality. (Furmans, 2008, p.80)	bad	ok	good	very good	0	
			Target achievement: The goals set by the management have to be reached within a certain time.	bad	ok	good	very good	0	
			Average					0.00	

Figure C.2: 1.1 System-CIP Execution (Furmans and Wlcek, 2012; Sobanski, 2009; Rother and Shook, 2008)

Warehouse Analysis 1.0			Level					Points	Comments
No.	Topics		0	1	2	3	4		
			Standards	Standards	Standards	Standards	Standards		
1.2	Point CIP	CONCEPT	**Target condition** Dehdari et al. (2011) mention that a target condition with a high maturity consist of a standard, a key performance indicator, and stability criteria. Additionally, standards should be visualized on the shop floor.	bad	ok	good	very good	0	
			Quick reaction system A quick reaction system is a standard that explains how a employee should react if unexpected things occurs and defines what the escalation limitis for reporting it to his supervisor. (Dehdari et al., 2011)	bad	ok	good	very good	0	
			Regular communication Regular communication with associates increases awareness of work plans, individual and departmental performance, goals, assignments, improvements, and changes. (Sobanski, 2008, p. 203). Dehdari et al. (2011) mention that the agenda, duration, and the focus of the communication should be defined.	bad	ok	good	very good	0	
			Sustainable problem solving Problem solving activities are organized into team-based functions. In a highly mature system, employees are empowered to, utilize, participate, initiate, and lead problem-solving activities autonomously, without significant management involvement. (Sobanski, 2008, p.200) Additionally, structured problem solving methodologies should be used to determine the root causes of problems as they arise. (Sobanski, 2008, p.204)	bad	ok	good	very good	0	
			Process confirmation Quality verification and inspection procedures in functions ensure that the standard operating procedures for each process are performed with minimal errors. (Sobanski, 2008, p205) The different hierarchy levels are also involved in the process confirmation. (Dehdari et al, 2011)	bad	ok	good	very good	0	
			Average					0.00	

Figure C.3: 1.2 Point-CIP Concept (Dehdari et al., 2011; Sobanski, 2009)

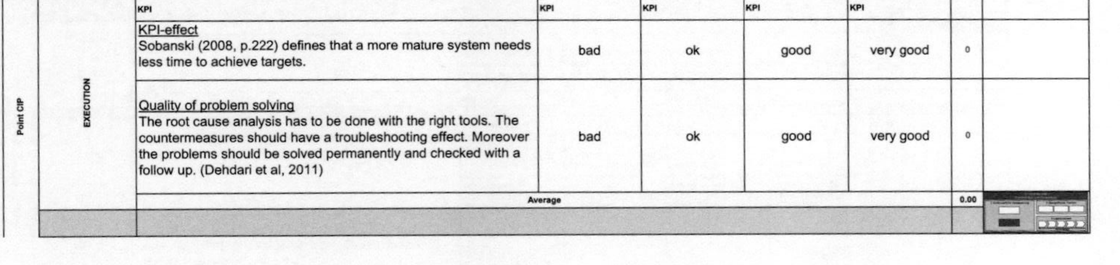

1.2	Point CIP	EXECUTION	KPI	KPI	KPI	KPI	KPI		
			KPI-effect Sobanski (2008, p.222) defines that a more mature system needs less time to achieve targets.	bad	ok	good	very good	0	
			Quality of problem solving The root cause analysis has to be done with the right tools. The countermeasures should have a troubleshooting effect. Moreover the problems should be solved permanently and checked with a follow up. (Dehdari et al, 2011)	bad	ok	good	very good	0	
				Average				0.00	

Figure C.4: 1.2 Point-CIP Execution (Sobanski, 2009; Dehdari et al., 2011)

Warehouse Analysis 1.0		Level					Points	Comments
No.	Topics	0	1	2	3	4		
		Standards	Standards	Standards	Standards	Standards		
2.1	Failure Prevention System — CONCEPT	**Work content** First, the failure has to detected. If measures or processes are installed that support detection and even prevention, then the failure prevention system is mature. (Hoyle, 2006, p.34) Parts identified as bad have to be taken right out of the process. (Vahrenkamp, 2010, p174)	bad	ok	good	very good	0	
			Visualization: - visualization to support failure detection implemented	Visualization: - visualization to support failure correction implemented (e. g. visualized reaction limit, visualized urgency plans)	Visualization: - visualization to support failure prevention (e.g. visualized instructions for packing etc.)	Visualization: - visualization of measured process parameters as an early warning system	0	
		Average					0.00	
	Failure Prevention System — EXECUTION	KPI	KPI	KPI	KPI	KPI		
			Internal error rate: - distinction between internally caused and external (non-owing) failures	Internal error rate: - stable achievement or positive trend of failure for more than 6 months	Internal error rate: - stable achievement or positive trend of failure for more than 1 year	Internal error rate: - stable achievement or positive trend of failure for more than 2 years	0	
		Average					0.00	

Figure C.5: 2.1 Failure Prevention System (Hoyle, 2006; Vahrenkamp, 2010)

Warehouse Analysis 1.0		Level					Points	Comments
No.	Topics	0	1	2	3	4		
		Standards	Standards	Standards	Standards	Standards		
2.2	Employee Involvement — CONCEPT	**Involvement:** Worker involvement is important for sustaining improvements and empowering the workforce. Thus, the percentage of the activities that are initiated by the workers is tracked. (Sobanski, 2008, p.168)	bad	ok	good	very good	0	
		Target deployment Warehouses targets have to be derived for each team. The team leader set the targets for his specific team derivived from his personal targets. (Jackson, 1996, p.109)	bad	ok	good	very good	0	
			Qualification / Training: - bottleneck treatment by flexible workforce	Qualification / Training: - regular training (off-the-job) for team leader provided to facilitate effective teaming, communication skills, continuous improvement methods (e. g. Methods Shop)	Qualification / Training: - regular training (off-the-job) for employees provided to facilitate effective teaming, communication skills, continuous improvement methods (e. g. Methods Shop)	Qualification / Training: - as level 3	0	
		Average					0.00	
	Employee Involvement — EXECUTION	KPI	KPI	KPI	KPI	KPI		
		Multi-skilled operators: Operators are not trained for different functions	Multi-skilled operators: 25-49 % of the operators are trained for more than one function	Multi-skilled operators: 50-74 % of the operators are trained for more than one function	Multi-skilled operators: 75-90 % of the operators are trained for more than one function	Multi-skilled operators: More than 90 % of the operators are trained for more than one function	0	
		Operator involvement Employees practice, exhibit the initiative, and adhere to the lean initiative, originating problem-solving and resolution activities individually and autonomously. (Sobanski, 2008, p. 201)	bad	ok	good	very good	0	
		Leadership involvement Manager or Supervisors spend significant time on the shop floor developing team leaders and employees and directing and facilitating daily activities. (Sobanski, 2008, p. 201)	bad	ok	good	very good	0	
		Average					0.00	

Figure C.6: 2.2 Employee Involvement (Jackson and Jones, 1996; Sobanski, 2009)

Warehouse Analysis 1.0			Level					Points	Comments
No.	Topics		0	1	2	3	4		
		Standards	Standards	Standards	Standards	Standards			
2.3	Standardized Work	CONCEPT	Coverage of standardized work The coverage can be checked by asking if there are current standardized worksheets for each major operation/process in each function. Smaller standardized work cycle lengths increase process resolution, bring problems to surface faster, reduce batch sizes, queuing, and WIP. (Sobanski, 2008, p. 194)	bad	ok	good	very good	0	
			Visualization Standardized worksheets has to be posted on the shop floor. (Sobanski, 2008, p. 194) They have to be also comprehensive and supported by visuals. (Graupp, 2006, p. 54)	bad	ok	good	very good	0	
			Qualification Employee understanding is increased by training and participation in continuous improvement of daily work activities. (Sobanski, 2008, p. 215)	bad	ok	good	very good	0	
			Average					0.00	
		KPI	KPI	KPI	KPI	KPI	KPI		
		EXECUTION	5S status 5S methodology for developing a place for everything and having everything in its place in the facility. (Sobanski, 2008, p. 217)	bad	ok	good	very good	0	
			Stability The percent of actual cycle counts performed daily versus department goals? Are they tracked and goals set? (Sobanski, 2008, p. 206)	bad	ok	good	very good	0	
			Productivity Productivity rates are tracked and displayed regularly versus facility and departmental goals? The actual productivity rates versus departmental and facility goals, where a higher ratio is better? (Sobanski, 2008, p. 212)	bad	ok	good	very good	0.00	
			Average						

Figure C.7: 2.3 Standardized Work (Sobanski, 2009; Graupp and Wrona, 2006)

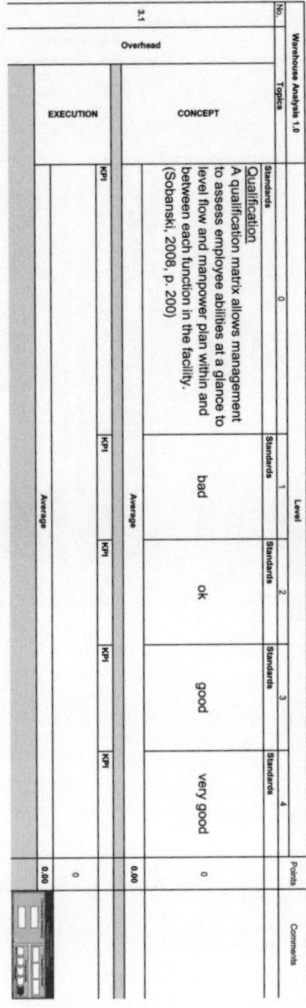

Warehouse Analysis 1.0		Level					Points	Comments
No.	Topics	0	1	2	3	4		
		Standards	Standards	Standards	Standards	Standards		
3.1	Overhead — CONCEPT	Qualification A qualification matrix allows management to assess employee abilities at a glance to level flow and manpower plan within and between each function in the facility. (Sobanski, 2008, p. 200)	bad	ok	good	very good	0	
		Average					0.00	
	EXECUTION	KPI	KPI	KPI	KPI	KPI	0	
		Average					0.00	

Figure C.8: 3.1 Overhead (Sobanski, 2009)

Warehouse Analysis 1.0			Level					Points	Comments
No.	Topics		0	1	2	3	4		
			Standards	Standards	Standards	Standards	Standards		
3.2	Outgoing Goods	CONCEPT	**Organization** Time windows can help level the workload for defined shipping times. If these time windows are also available for the downstream processes, the maturity is higher. (Dehdari et al, 2012)	bad	ok	good	very good	0	
				Technical Equipment: - OG operator has to handle the entire process manually	Technical Equipment: - OG operator its supported by semi-automatic equipment (e. g. scanner)	Technical Equipment: - as level 3	Technical Equipment: - OG operator is supported by automatic equipment (e. g. RFID gate)	0	
			Visualization Visual controls can help guarantee the time windows. Visual control mechanisms enhance process integrity and reduce waste by eliminating searching and stabilizing processes. (Sobanski, 2008, p.215) (Furmans, 2012, p.49)	bad	ok	good	very good	0	
								0	
			Average					0.00	
			KPI	KPI	KPI	KPI	KPI		
		EXECUTION	**Time window adherence** Tracking the time window adherence information illustrates the performance versus the expectations. (Sobanski, 2008, p. 212)	bad	ok	good	very good	0	
			Balancing of complete shipping processes To balance the workload, it is important to know the workload and available man hours. An improvement of the balance is desired. (Furmans, 2012, p. 82)	bad	ok	good	very good	0	
			Lead time of the shipping preparation process The lead time of the shipping and shipping preparation should be measured. If it is stable reduced for a certain time than this is a good indication. (Furmans, 2012, p. 79)	bad	ok	good	very good	0	
				Dispatch error rate: - failures caused in the dispatch area are measured	Dispatch error rate: - stable achievement or positive trend of failure for more than 6 months	Dispatch error rate: - stable achievement or positive trend of failure for more than 1 year	Dispatch error rate: - stable achievement or positive trend of failure for more than 2 years	0	
				Handling steps: - handling steps are counted	Handling steps: - reduction in handling steps in the last 6 months by optimizing measures (if necessary minimum is reached, stable achievement counts as reduction)	Handling steps: - reduction in handling steps in the last 3 months by optimizing measures (if necessary minimum is reached, stable achievement counts as reduction)	Handling steps: - reduction in handling steps in the last month by optimizing measures (if necessary minimum is reached, stable achievement counts as reduction)	0	
			Average					0.00	

Figure C.9: 3.2 Outgoing Goods (Dehdari and Schwab, 2012; Sobanski, 2009; Furmans and Wlcek, 2012)

Warehouse Analysis 1.0			Level								Points	Comments
No.	Topics		0		1		2		3		4	
			Standards		Standards		Standards		Standards		Standards		
3.3	Packaging	CONCEPT			Packaged good: - packaged goods are available in short walking distance (< 5 meters)	Packaged good: - packaged goods are available for the packaging operator in operating distance (without walking)	Packaged good: - packaged goods are delivered in the right sequence (regarding dispatch sequence)	Packaged good: - as level 3	0	
					Packaging material: - operator has packaging material in walking distance - material must be restocked by packaging operator himself (interruption of work)	Packaging material: - Operator has packaging material in walking distance - Material is restocked regularly	Packaging material: - operator has packaging material in operating distance (without walking) - material is restocked regularly - operator has possibility for emergency call in case material runs out	Packaging material: - as level 3 - staging of material is guaranteed by a well-working, consumption-driven system (e. g. milkrun)	0	
					Packaging process: - ergonomic requirements are fulfilled regarding packaging workplace	Packaging process: - process-oriented workplace design/layout: central and sufficient disposal of all devices / tools / machines / IT-support	Packaging process: - as level 2 - flexible workplace design --> worksplace can be used for different functions - standards for packaging process are defined and communicated	Packaging process: - operators processing a packaging unit are working > 90 % of their work time according to standardized work	0	
					Visualization: - clear visualization of which products must be packed together - standardized and visualized place where to put the delivery note	Visualization: - as level 1 - clear visualization of which consigned orders should be worked on next (FIFO, rush orders)	Visualization: - as level 2 - easily visible completeness check (e. g. by weight)	Visualization: - as level 3 - visualization, if packaging operators are still on time or if consignmnt operators are working faster --> flexible employee change - based on volume calculation of order	0	
							Average				0.00	
		KPI			KPI	KPI	KPI	KPI				
		EXECUTION			Lead time of the packaging process: - lead time of the packaging process is measured	Lead time of the packaging process: - positive trend or at stabie target level of average lead time of all packing units since min. 6 months	Lead time of the packaging process: - positive trend of average lead time since the last 12 months or stable	Lead time of the packaging process: - positive trend of average lead time since the last 24 months or on outstanding level	0	
					Packaging error rate: - failures caused in the packaging area are measured	Packaging error rate: - stable achievement or positive trend of failure for more than 6 months	Packaging error rate: - stable achievement or positive trend of failure for more than 1 year	Packaging error rate: - stable achievement or positive trend of failure for more than 2 years	0	
							Average				0.00	

Figure C.10: 3.3 Packaging

Warehouse Analysis 1.0		Level						Points	Comments
No.	Topics		0	1	2	3	4		
3.4	Picking	**CONCEPT**	**Standards**	**Standards** Picking process: - all necessary MAE are available sufficiently - consigment operators are supported by manual measures (e.g. picklist) - picking workplaces fulfill ergonomic requirements	**Standards** Picking process: - as level 1 - picking operators are supported by systematic and automatic measures (e.g. pick-by-scan) - operator self control - picking workplaces are process oriented	**Standards** Picking process: - as level 2 - picking operators are supported by systematic and automatic measures that allow two free hands (e.g. pick by voice/light)	**Standards** Picking process: - as level 3 - operators processing a packing unit are working >90% of their work time according to standardized work	0	
				Organizational system: - defined process for treatment of fix date and rush order exists	Organizational system: - picking technique depends on order frequency, order volume/size, stability of goods, packaging material - system calculates optimal method - job control follows a clearly defined strategy	Organizational system: - as level 2 - picking sequence is optimized by system - order dispatch depends on available picking performance, current working progress and current system status	Organizational system: - as level 3 - picking racks are flexible - systematic and extensive planning approach to level capacity short term in case of fluctuation in demand	0	
				Information system: - manual order entry	Information system: - failure prevention by automatic registration of picking positions (e.g. by scanning)	Information system: - automatical order entry - voucherless transfer of picking order	Information system: - as level 3	0	
				Visualization: - visualization to support material flow system (transport, taking, hand-over), e. g. picking areas, ...	Visualization: - visualization to support organizational system (order sequence, picking amount, internal delivery performance, ...)	Visualization: - as level 2	Visualization: - as level 3	0	
				Average				0.00	
		EXECUTION	**KPI**	**KPI** Lead time of the picking process: - lead time of the picking process is measured	**KPI** Lead time of the picking process: - positive trend or at stable target level of average lead time of all packaging units since min. 6 months	**KPI** Lead time of the picking process: - positive trend of average lead time since the last 12 months or stable	**KPI** Lead time of the picking process: - positive trend of average lead time since the last 24 months or on outstanding level	0	
				Picking error rate: - failures caused in the picking area are measured	Picking error rate: - stable achievement or positive trend of failure for more than 6 months	Picking error rate: - stable achievement or positive trend of failure for more than 1 year	Picking error rate: - stable achievement or positive trend of failure for more than 2 years	0	
				Average				0.00	

Figure C.11: 3.4 Picking

Warehouse Analysis 1.0			Level					Points	Comments
No.	Topics		0	1	2	3	4		
			Standards	**Standards**	**Standards**	**Standards**	**Standards**		
3.5	Storage	CONCEPT	Storage technic/layout: - safety-related and judicial guidelines are considered	Storage technic/layout: - as level 0 - storage technique depends on goods (size, weight, packaging, stability, amount, ...)	Storage technic/layout: - as level 1 - storage technique depends on economic factors (optimal land and room use)	Storage technic/layout: - as level 2 - storage technic is updated min. 2x/year - automatization level fits to aimed profitability and flexibility (can be argumented by warehouse manager)	Storage technic/layout: - as level 2 - storage technic is updated min. 6x/year - as level 3	0	
			Storage criteria: - for storage safety-related and judicial guidelines are considered (especially in case of hazardous materials)	Storage criteria: - as level 0 - technical requirements are considered (optimal volume use, equal load, ...)	Storage criteria: - as level 1 - economic facts are considered (inventory turnover, availability, optimal distances ...)	Storage criteria: - technical and economic facts are updated min. 2x/year	Storage criteria: - technical and economic facts are uptadet min. 6x/year	0	
				Inventory management: - clear and systematic administration of all inventories in the warehouse system	Inventory management: - as level 1 - little/no time window between moving goods and booking process	Inventory management: - as level 2 - permanent inventory taking by highly developed EDV systems	Inventory management: - as storage process doesn't allow any inventory deviations	0	
				Visualization: - bin locations are clearly visualized (e.g. number, scan code)	Visualization: - as level 1 - min/max inventories are visualized (if existing)	Visualization: - as level 2 - inventories and performance of the warehouse are visible for everyone (e. g. screen with current KPIs)	Visualization: - as level 3	0	
			In case storage is seperated from consignment area					0	
			From-bin transfer:	From-bin transfer: - Transfer to consignment area is consumption driven (Pull)	From-bin transfer: - as level 2	From-bin transfer: - as level 3		0	
			Average					0.00	
		EXECUTION	**KPI**	**KPI**	**KPI**	**KPI**	**KPI**		
				Lead time of the storage process: - lead time of the storage process is measured	Lead time of the storage process: - positive trend or at stable target level of average lead time of all packaging units since min. 6 months	Lead time of the storage process: - positive trend of average lead time since the last 12 months or stable	Lead time of the storage process: - positive trend of average lead time since the last 24 months or on outstanding level	0	
				Storage error rate: - failures caused in the area of storage are measured	Storage error rate: - stable achievement or positive trend of failure for more than 6 months	Storage error rate: - stable achievement or positive trend of failure for more than 1 year	Storage error rate: - stable achievement or positive trend of failure for more than 2 years	0	
			Average					0.00	

Figure C.12: 3.5 Storage

Warehouse Analysis 1.0		Level					Points	Comments
No.	Topics	0	1	2	3	4		
		Standards	Standards	Standards	Standards	Standards		
3.6	Incoming Goods	CONCEPT	Organization Time windows can help level the workload for defined receiving times. If these time windows are also available for the upstream processes, the maturity is higher. (Dehdari et al, 2012)	bad	ok	good	very good	0
				Entry / booking: - two-level Entry: the posting of the goods to the inventory and the entry of the incoming goods are done seperately	Entry / booking: - as level 1 - entry in the system is done not more than one day after goods are received	Entry / booking: - single-level Entry: the posting of the goods to the inventory is done together with the entry of the incoming goods (at least interim booking of Avis-Information to storage system = Einbuchung unter Vorbehalt)	Entry / booking: - as level 3 - entry in the system is done nearly at the same time as goods are received (not more than 2 hours later)	0
				Inspection: - inspection intensity depends on failure rate of supplier or comparable criteria	Inspection: - supplier deliver already filled out and signed control documents with their goods	Inspection: entry in the system is - free ticket (Freipass) for audited suppliers (e. g. skip-quota after a defined scheme)	Inspection: - as level 3	0
			Visualization Visual controls can help guarantee the time windows. Visual control mechanisms enhance process integrity and reduce waste by eliminating searching and stabilizing processes. (Sobanski, 2008, p.215) (Furmans, 2012, p.49)	bad	ok	good	very good	0
			Average					0.00

Figure C.13: 3.6 Incoming Goods Concept (Dehdari and Schwab, 2012; Furmans and Wlcek, 2012)

3.6	Incoming Goods	EXECUTION							
			KPI	KPI	KPI	KPI	KPI		
			Time window adherence Tracking the time window adherence information illustrates the performance versus the expectations. (Sobanski, 2008, p. 212)	bad	ok	good	very good	0	
			Balancing of complete incoming processes To balance the workload, it is important to know the workload and available man hours. An improvement of the balance is desired. (Furmans, 2012, p. 82)	bad	ok	good	very good	0	
			Lead time of the incoming process The lead time of the incoming and incoming handling should be measured. If it is stabile reduced for a certain time than this is a good indication. (Furmans, 2012, p. 79)	bad	ok	good	very good	0	
				Receiving error rate: - failures caused in the area of incoming goods are measured	Receiving error rate: - stable achievement or positive trend of failure for more than 6 months	Receiving error rate: - stable achievement or positive trend of failure for more than 1 year	Receiving error rate: - stable achievement or positive trend of failure for more than 2 years	0	
				Handling steps - handling steps are counted	Handling steps: - reduction of handling steps in the last 6 months by optimizing measures (if necessary minimum is reached, stable achievement counts as reduction)	Handling steps: - reduction of handling steps in the last 3 months by optimizing measures (if necessary minimum is reached, stable achievement counts as reduction)	Handling steps: - reduction of handling steps in the last month by optimizing measures (if necessary minimum is reached, stable achievement counts as reduction)	0	
		Average						0.00	

Figure C.14: 3.6 Incoming Goods Excecution (Furmans and Wlcek, 2012; Sobanski, 2009)

D Appendix – Warehouse Excellence Group Assessment Results

| 2010 | | | | | | | | | | | | | | | | | | | |
Criteria	Min	Max	Average	W1	W2	W3	W4	W5	W6	W7	W8	W9	W10	W11	W12	W13	W14	W15	W16
1.1 System-CIP Concept																			
Business Requirements	0	4	0.750	0	2	0	0	0	4	0	0	2	0	2	0	2	0	0	0
Value Stream Planning	0	1	0.375	0	1	1	0	0	1	1	0	0	0	1	1	0	0	0	0
Identification of Improvement Activities	0	1	0.125	0	1	0	0	0	1	0	0	0	0	0	0	0	0	0	0
Definition of Target Conditions	0	2	0.125	0	2	0	0	0	0	0	0	0	0	0	0	0	0	0	0
System-CIP Projects	0	1	0.188	0	1	0	0	0	1	0	0	0	1	0	0	0	0	0	0
Point-CIP Areas	0	4	1.125	0	4	1	2	0	1	1	0	1	1	1	0	1	1	4	0
1.1 System-CIP Execution																			
Target Derivation	0	1	0.688	0	1	1	1	0	1	1	1	1	1	1	0	1	0	1	0
System-CIP Cycles	0	1	0.250	0	1	0	0	0	1	0	0	1	0	1	0	0	0	0	0
Improvement Focus	0	2	0.938	0	0	2	1	1	1	1	1	1	1	2	0	1	2	0	1
Leadership Involvement	0	2	0.313	0	1	0	0	0	2	0	0	0	0	2	0	0	0	0	0
VSM-Quality	0	4	0.750	0	1	4	0	0	4	1	0	0	0	1	1	0	0	0	0
Target Achievement	0	2	0.125	0	0	0	0	0	0	0	0	0	0	0	0	2	0	0	0
1.2 Point-CIP Concept																			
Target Condition	0	2	0.438	0	2	1	0	0	1	1	0	0	1	1	0	0	0	0	0
Quick Reaction System	0	4	0.438	0	0	0	0	0	2	0	0	1	0	4	0	0	0	0	0
Regular Communication	0	3	1.375	1	2	1	1	1	2	1	0	2	1	2	0	1	3	3	1
Sustainable Problem Solving	0	2	0.438	0	0	0	0	0	2	0	0	0	0	2	0	1	0	2	0
Process Confirmation	0	2	0.313	0	0	0	0	0	1	0	0	1	0	0	0	0	1	2	0
1.2 Point-CIP Execution																			
KPI-Effect	0	3	0.188	0	0	0	0	0	0	0	0	0	0	3	0	0	0	0	0
Quality of Problem Solving	0	2	0.250	0	0	0	0	0	2	0	0	0	0	2	0	0	0	0	0
Sum of Reached Maturity Level				1	19	11	5	2	27	7	2	10	6	25	2	9	7	12	2
Sum of Average			9.188																
Standard Deviation			8.109																
Coefficient of Variation			88.266																

Table D.1: BLWA results per warehouse: Warehouse Excellence group 2010

2011																		
Criteria	Min	Max	Average	W1	W2	W3	W4	W5	W6	W7	W8	W9	W10	W11	W12	W13	W14	W15	W16
1.1 System-CIP Concept																			
Business Requirements	1	4	2.000	1	4	2	1	2	4	1	1	4	1	1	1	4	2	2	1
Value Stream Planning	0	4	1.875	1	2	2	2	3	4	1	1	3	2	3	1	1	0	3	1
Identification of Improvement Activities	0	4	1.375	2	2	1	1	4	2	0	0	4	0	1	0	0	1	4	0
Definition of Target Conditions	0	3	1.438	1	3	1	2	1	2	2	2	1	2	1	1	1	0	2	1
System-CIP Projects	0	2	1.375	1	2	1	0	1	2	2	1	1	2	2	1	2	0	2	2
Point-CIP Areas	1	4	2.125	2	4	2	2	2	3	2	1	2	2	1	1	2	2	4	2
1.1 System-CIP Execution																			
Target Derivation	1	4	1.375	1	4	1	1	1	1	1	1	4	1	1	1	1	1	1	1
System-CIP Cycles	0	2	1.000	1	2	1	1	1	1	1	1	1	1	1	0	1	0	1	2
Improvement Focus	1	4	2.563	2	2	2	4	3	4	1	1	2	4	2	2	4	2	4	2
Leadership Involvement	0	4	2.188	1	2	0	0	4	4	2	2	4	2	4	0	2	0	4	4
VSM-Quality	1	4	1.938	1	1	4	1	1	4	1	1	4	1	1	1	1	1	4	4
Target Achievement	0	3	1.688	1	1	0	3	2	1	2	1	3	2	1	3	3	0	2	2
1.2 Point-CIP Concept																			
Target Condition	1	2	1.438	1	2	1	1	1	1	2	2	1	2	1	1	2	1	2	2
Quick Reaction System	0	4	3.250	4	2	1	4	4	4	1	4	4	0	4	4	4	4	4	4
Regular Communication	1	4	2.813	4	2	1	4	3	2	3	2	3	4	2	3	2	3	3	4
Sustainable Problem Solving	0	3	1.313	0	2	0	3	3	3	1	0	0	0	2	2	1	0	2	2
Process Confirmation	0	2	1.063	1	1	2	1	0	2	0	1	1	1	1	1	0	1	2	2
1.2 Point-CIP Execution																			
KPI-Effect	0	3	1.563	0	1	0	2	2	1	3	1	3	2	1	3	1	0	2	3
Quality of Problem Solving	0	2	0.438	0	1	0	2	0	2	0	0	0	0	0	1	1	0	0	0
Sum of Reached Maturity Level				25	40	22	35	38	47	26	23	45	29	30	27	33	18	48	39
Sum of Average			32.813																
Standard Deviation			9.304																
Coefficient of Variation			28.355																

Table D.2: BLWA results per warehouse: Warehouse Excellence group 2011

E Appendix – Warehouse Excellence Group KPR

Description	2010 Lean-Index	2011 Lean-Index	Delta	Slope	2010 Average	2011 Average	Jan 10	Feb 10	Mar 10	Apr 10	May 10	Jun 10	Jul 10	Aug 10	Sep 10	Oct 10	Nov 10	Dec 10	
W1	1.000	25.000	21.000	0.002	N/A	110.029	N/A	N/A	N/A	N/A	N/A	N/A	N/A	N/A	N/A	N/A	N/A	N/A	...
W2	19.000	40.000	20.000	0.023	126.279	136.333	100.000	119.359	118.073	126.857	129.741	124.416	125.169	120.653	135.050	135.098	143.011	137.917	...
W3	11.000	22.000	7.000	0.018	100.522	106.876	100.000	103.543	98.057	94.107	108.449	103.961	102.367	105.836	93.689	95.866	98.265	102.120	...
W4	5.000	35.000	30.000	0.013	103.876	110.017	100.000	94.208	112.149	117.247	104.018	105.892	98.844	98.199	104.154	104.867	107.941	98.990	...
W5	2.000	38.000	36.000	0.076	107.323	136.543	100.000	103.012	110.100	103.789	98.115	99.963	104.624	104.005	114.530	111.794	112.779	125.158	...
W6	27.000	47.000	16.000	0.055	109.987	135.913	100.000	110.879	103.724	116.915	111.403	109.172	107.776	119.968	103.406	105.370	104.260	126.971	...
W7	7.000	26.000	19.000	0.010	109.088	114.664	100.000	105.941	110.839	120.476	109.851	109.957	109.957	109.957	108.294	114.353	99.478	...	...
W8	2.000	23.000	21.000	0.045	92.803	111.874	100.000	89.745	90.284	89.599	92.321	96.060	86.611	92.491	101.043	99.378	93.101	83.008	...
W9	10.000	45.000	36.000	0.047	100.000	123.865	100.000	100.000	100.000	100.000	100.000	100.000	100.000	100.000	100.000	100.000	100.000	100.000	...
W10	6.000	29.000	23.000	0.011	104.618	106.438	100.000	95.059	101.267	103.184	110.584	110.306	101.609	113.607	110.030	101.730	109.019	99.019	...
W11	25.000	30.000	3.000	0.035	96.724	107.438	100.000	98.656	85.541	98.037	99.141	97.299	92.516	77.927	100.022	98.020	114.107	99.417	...
W12	2.000	27.000	25.000	-0.021	98.543	90.645	100.000	95.678	95.678	99.804	103.733	100.982	104.322	103.340	96.071	100.000	98.035	84.872	...
W13	9.000	33.000	23.000	0.013	99.081	104.704	100.000	103.166	104.127	98.783	97.565	90.230	91.251	99.524	100.145	102.789	102.737	98.650	...
W14	7.000	18.000	10.000	0.214	108.948	194.287	100.000	102.511	106.540	104.389	100.880	107.706	99.690	135.415	119.667	117.291	110.206	103.083	...
W15	12.000	48.000	38.000	0.001	97.827	100.702	100.000	95.074	106.828	97.519	107.210	106.909	84.092	79.620	107.510	104.368	106.904	77.891	...
W16	2.000	39.000	37.000	0.001	104.353	105.146	100.000	106.149	107.574	103.754	108.967	100.951	102.366	113.170	105.651	106.348	107.729	89.572	...
Average	9.188	32.813	22.813	0.035	103.998	118.467	100.000	101.532	103.385	104.964	105.465	104.254	100.746	104.914	106.728	106.081	108.163	101.743	...
Standard Deviation	8.109	9.304																	

Table E.1: Warehouse Excellence Group KPR development 2010

Description		Jan 11	Feb 11	Mar 11	Apr 11	May 11	Jun 11	Jul 11	Aug 11	Sep 11	Oct 11	Nov 11	Dec 11
W1	...	100.000	113.145	105.996	113.920	107.362	123.234	103.032	121.913	117.068	105.090	102.782	106.806
W2	...	134.006	153.294	151.558	158.436	144.674	139.184	117.316	113.959	126.342	115.271	157.896	124.063
W3	...	96.552	102.014	102.683	105.324	107.971	102.820	105.616	116.740	107.548	108.633	109.956	116.657
W4	...	106.252	107.767	113.661	105.939	114.815	107.412	112.161	111.059	108.709	106.301	110.725	115.405
W5	...	127.533	130.672	134.933	129.112	127.030	118.556	140.215	142.221	143.088	143.241	150.185	151.728
W6	...	132.198	135.975	142.493	139.101	137.188	142.771	134.470	126.133	139.254	131.763	131.004	138.604
W7	...	113.997	120.000	111.113	116.220	122.342	113.416	117.804	111.965	108.123	112.420	115.502	113.070
W8	...	103.998	102.264	113.347	101.861	104.892	119.485	103.785	113.848	118.883	122.889	119.411	117.829
W9	...	128.902	138.158	132.555	117.495	124.410	122.193	107.952	108.377	109.437	131.758	134.954	130.186
W10	...	99.092	97.336	99.952	101.364	107.231	106.816	114.855	114.352	108.899	108.464	111.428	107.466
W11	...	95.066	93.509	97.092	98.719	85.117	116.911	114.831	112.400	120.904	133.292	117.118	104.295
W12	...	89.980	90.766	92.338	101.965	96.464	87.220	86.163	94.636	92.010	86.745	90.022	79.434
W13	...	104.051	100.322	105.792	104.792	100.770	105.221	105.088	107.114	102.284	104.442	106.182	110.391
W14	...	121.168	126.660	117.900	195.811	219.883	221.813	186.839	337.651	215.681	209.188	211.305	167.545
W15	...	101.126	105.547	103.660	94.869	111.113	102.375	85.357	82.176	121.011	107.383	116.488	77.324
W16	...	102.370	101.258	108.785	92.102	112.674	108.821	105.435	104.782	115.269	110.481	105.388	94.390
Average	...	109.768	113.668	114.616	117.314	120.246	121.140	115.057	126.208	122.157	121.085	124.397	115.949

Table E.2: Warehouse Excellence KPR development 2011

F Appendix – Control Group Assessment Results

2010 Criteria	Min	Max	Average	C1	C2	C3	C4	C5	C6	C7	C8	C9	C10	C11	C12	C13	C14	C15	C16	C17	C18	C19	C20	C21	C22	C23	C24	C25	C26	C27	C28
1.1 System-CIP Concept																															
Business Requirements	0	2	0.286	0	0	0	1	1	0	1	2	0	0	0	0	0	0	0	0	0	0	0	0	0	0	0	0	0	0	2	1
Value Stream Planning	0	1	0.036	1	0	0	0	0	0	0	0	0	0	0	0	0	0	0	0	0	0	0	0	0	0	0	0	0	0	0	0
Identification of Improvement Activities	0	0	0.000	0	0	0	0	0	0	0	0	0	0	0	0	0	0	0	0	0	0	0	0	0	0	0	0	0	0	0	0
Definition of Target Conditions	0	2	0.250	0	0	0	2	0	2	1	0	0	0	0	0	0	0	0	0	0	0	0	0	0	0	0	0	0	0	1	1
System-CIP Projects	0	0	0.000	0	0	0	0	0	0	0	0	0	0	0	0	0	0	0	0	0	0	0	0	0	0	0	0	0	0	0	0
Point-CIP Areas	0	3	0.429	3	1	0	1	1	2	3	0	0	0	0	0	0	0	0	0	0	0	0	0	0	0	0	0	0	0	1	0
1.1 System-CIP Execution																															
Target Derivation	0	4	0.464	0	0	1	1	1	4	0	4	0	0	0	0	0	0	0	0	0	0	0	0	0	0	0	0	0	0	1	1
System-CIP Cycles	0	0	0.000	0	0	0	0	0	0	0	0	0	0	0	0	0	0	0	0	0	0	0	0	0	0	0	0	0	0	0	0
Improvement Focus	0	4	0.714	2	0	3	4	4	1	0	1	0	0	0	0	0	0	0	0	0	0	0	0	0	0	0	0	0	0	2	3
Leadership Involvement	0	0	0.000	0	0	0	0	0	0	0	0	0	0	0	0	0	0	0	0	0	0	0	0	0	0	0	0	0	0	0	0
VSM-Quality	0	0	0.000	0	0	0	0	0	0	0	0	0	0	0	0	0	0	0	0	0	0	0	0	0	0	0	0	0	0	0	0
Target Achievement	0	0	0.000	0	0	0	0	0	0	0	0	0	0	0	0	0	0	0	0	0	0	0	0	0	0	0	0	0	0	0	0
1.2 Point-CIP Concept																															
Target Condition	0	2	0.357	2	0	0	2	2	0	1	0	0	0	0	0	0	0	0	0	0	0	0	0	0	0	0	0	0	0	2	1
Quick Reaction System	0	2	0.250	2	0	0	2	2	0	1	0	0	0	0	0	0	0	0	0	0	0	0	0	0	0	0	0	0	0	0	0
Regular Communication	0	4	0.750	3	3	0	1	2	4	0	0	0	0	0	4	0	0	0	0	0	0	0	0	0	0	0	0	0	0	1	3
Sustainable Problem Solving	0	4	0.750	3	3	0	0	0	0	4	0	0	0	0	4	0	0	0	0	0	0	0	0	0	0	0	0	0	0	4	3
Process Confirmation	0	2	0.179	2	1	0	0	0	0	1	0	0	0	0	0	0	0	0	0	0	0	0	0	0	0	0	0	0	0	1	0
1.2 Point-CIP Execution																															
KPI-Effect	0	2	0.250	0	0	2	2	2	0	1	0	0	0	0	0	0	0	0	0	0	0	0	0	0	0	0	0	0	0	0	0
Quality of Problem Solving	0	4	0.750	2	2	1	3	3	3	1	0	0	0	0	0	0	0	0	0	0	0	0	0	0	0	0	0	0	0	2	4
Sum of Reached Maturity Level				20	10	7	19	18	16	14	7	0	0	0	8	0	0	0	0	0	0	0	0	0	0	0	0	0	0	17	17
Sum of Average C1-C56			5.464																												
Standard Deviation of Sum C1-C56			7.488																												
Coefficient of Variation C1-C56			137.028																												

Table F.1: Assessment results per warehouse: control group 2010 C1-C28

2010 Criteria	C29	C30	C31	C32	C33	C34	C35	C36	C37	C38	C39	C40	C41	C42	C43	C44	C45	C46	C47	C48	C49	C50	C51	C52	C53	C54	C55	C56
1.1 System-CIP Concept																												
Business Requirements	0	2	2	0	1	0	0	2	2	1	2	0	0	2	2	2	1	0	2	2	0	0	0	0	2	0	2	0
Value Stream Planning	0	0	0	0	2	0	0	0	0	1	1	0	0	0	0	0	0	0	1	0	0	0	0	0	0	0	1	0
Identification of Improvement Activities	0	0	0	0	0	0	0	0	0	0	0	0	0	0	0	0	0	0	4	0	0	0	0	0	0	0	0	0
Definition of Target Conditions	2	0	0	2	0	0	0	4	0	0	0	0	1	0	0	2	1	1	0	4	0	0	1	0	0	0	0	0
System-CIP Projects	0	0	0	0	0	0	0	0	0	4	0	0	0	0	0	0	0	0	0	0	0	0	0	0	0	0	0	0
Point-CIP Areas	0	0	0	0	1	0	0	3	0	2	0	2	1	0	0	0	1	1	1	3	1	0	3	0	0	0	2	1
1.1 System-CIP Execution																												
Target Derivation	0	1	0	0	0	0	0	1	0	0	0	0	1	0	0	1	0	0	0	4	1	0	1	0	0	0	0	0
System-CIP Cycles	0	0	0	0	1	0	0	0	0	1	0	0	0	0	0	0	0	0	0	0	0	0	0	0	0	0	0	0
Improvement Focus	0	2	0	4	4	4	0	2	4	2	4	1	0	4	4	2	2	3	1	2	2	0	0	0	0	0	0	1
Leadership Involvement	0	0	0	0	0	0	0	0	0	0	0	0	0	0	0	0	0	0	1	0	0	0	0	0	0	0	1	0
VSM-Quality	0	0	0	0	0	0	0	0	0	0	0	0	0	0	0	0	0	0	0	0	0	0	0	0	0	0	1	0
Target Achievement	0	0	0	0	1	0	0	0	0	1	0	0	0	0	0	0	0	0	0	0	0	0	0	0	0	0	0	0
1.2 Point-CIP Concept																												
Target Condition	1	2	0	1	0	1	1	2	1	2	1	2	2	1	1	1	1	1	0	0	2	0	0	0	2	0	0	0
Quick Reaction System	0	0	0	0	0	2	0	2	2	0	2	0	2	2	2	0	0	0	2	4	0	0	4	0	0	0	0	0
Regular Communication	2	4	0	0	3	0	2	0	4	3	1	1	1	1	1	1	0	0	0	0	4	0	1	0	0	0	2	1
Sustainable Problem Solving	4	3	0	0	1	0	0	2	0	0	0	1	4	0	0	0	3	0	0	4	0	0	4	0	3	0	4	3
Process Confirmation	1	0	0	0	2	0	0	0	0	2	0	0	2	0	0	1	0	2	2	0	0	0	0	0	0	0	0	0
1.2 Point-CIP Execution																												
KPI-Effect	0	0	0	0	4	0	0	0	0	2	0	0	0	0	0	0	1	1	1	1	1	0	0	0	0	0	0	0
Quality of Problem Solving	4	4	0	1	0	1	4	1	1	2	1	0	3	1	1	1	2	0	1	4	1	0	2	0	3	0	4	1
Sum of Reached Maturity Level	14	18	2	8	20	8	7	19	14	23	12	7	17	11	11	11	12	9	16	28	12	0	16	0	10	0	17	7

Table F.2: Assessment results per warehouse: control group 2010 C29-C56

2011																																
Criteria	Min	Max	Average	C1	C2	C3	C4	C5	C6	C7	C8	C9	C10	C11	C12	C13	C14	C15	C16	C17	C18	C19	C20	C21	C22	C23	C24	C25	C26	C27	C28	…
1.1 System-CIP Concept																																
Business Requirements	0	3	0.821	1	0	2	1	3	2	1	2	0	0	0	0	0	0	0	0	0	0	0	0	0	0	0	0	0	0	1	1	…
Value Stream Planning	0	4	0.482	1	0	0	3	4	0	0	0	0	0	0	0	0	0	0	0	0	0	0	0	0	0	0	0	0	0	0	0	
Identification of Improvement Activities	0	4	0.339	1	0	0	4	4	0	0	0	0	0	0	0	0	0	0	0	0	0	0	0	0	0	0	0	0	0	0	0	…
Definition of Target Conditions	0	4	0.732	2	0	1	4	4	1	1	0	0	1	0	0	0	0	0	0	0	0	0	0	0	0	0	0	0	0	0	0	…
System-CIP Projects	0	4	0.696	0	0	0	3	4	0	0	0	0	0	0	0	0	0	0	0	0	0	0	0	0	0	0	0	0	0	0	0	…
Point-CIP Areas	0	3	0.786	3	0	0	2	3	2	3	0	0	3	0	0	0	0	0	0	0	0	0	0	0	0	0	0	0	0	0	0	
1.1 System-CIP Execution																																
Target Derivation	0	4	0.339	0	0	0	4	4	0	0	0	0	1	0	0	0	0	0	0	0	0	0	0	0	0	0	0	0	0	1	0	…
System-CIP Cycles	0	4	0.286	0	0	0	1	1	0	0	0	0	0	0	0	0	0	0	0	0	0	0	0	0	0	0	0	0	0	0	0	…
Improvement Focus	0	4	0.893	3	0	0	3	1	1	0	1	0	1	0	0	0	0	0	0	0	0	0	0	0	0	0	0	0	0	4	0	…
Leadership Involvement	0	4	0.446	0	0	0	1	4	0	0	0	0	0	0	0	0	0	0	0	0	0	0	0	0	0	0	0	0	0	0	0	…
VSM-Quality	0	4	0.304	2	0	0	0	4	0	0	0	0	0	0	0	0	0	0	0	0	0	0	0	0	0	0	0	0	0	0	0	…
Target Achievement	0	2	0.071	0	0	0	1	1	0	0	0	0	0	0	0	0	0	0	0	0	0	0	0	0	0	0	0	0	0	0	0	
1.2 Point-CIP Concept																																
Target Condition	0	4	0.839	1	0	2	3	4	1	1	0	0	2	0	0	0	0	0	0	0	0	0	0	0	0	0	0	0	0	0	1	…
Quick Reaction System	0	4	0.946	0	0	2	4	4	4	1	4	0	4	0	2	0	0	0	0	0	0	0	0	0	0	0	0	0	0	0	0	…
Regular Communication	0	4	1.143	1	3	0	4	1	4	0	4	0	1	0	4	0	0	0	0	0	0	0	0	0	0	0	0	0	0	0	0	…
Sustainable Problem Solving	0	4	1.411	2	0	0	4	4	3	4	4	0	0	0	1	0	0	0	0	0	0	0	0	0	0	0	0	0	0	2	4	…
Process Confirmation	0	2	0.446	1	0	1	1	1	1	1	0	0	1	0	1	0	0	0	0	0	0	0	0	0	0	0	0	0	0	0	0	
1.2 Point-CIP Execution																																
KPI-effect	0	4	0.661	3	0	2	2	1	2	1	0	0	0	0	0	0	0	0	0	0	0	0	0	0	0	0	0	0	0	0	0	…
Quality of problem solving	0	4	1.179	3	0	3	4	4	3	1	1	0	1	0	3	0	0	0	0	0	0	0	0	0	0	0	0	0	0	1	0	…
Sum of Reached Maturity Level				24	3	13	49	56	24	14	16	0	15	0	11	0	0	0	0	0	0	0	0	0	0	0	0	0	0	9	6	…
Sum of Average C1-C56			12.821																													
Standard Deviation of Sum C1-C56			13.945																													
Coefficient of Variation C1-C56			108.767																													

Table F.3: Assessment results per warehouse: control group 2011 C1-C28

2011 Criteria	...	C29	C30	C31	C32	C33	C34	C35	C36	C37	C38	C39	C40	C41	C42	C43	C44	C45	C46	C47	C48	C49	C50	C51	C52	C53	C54	C55	C56
1.1 System-CIP Concept																													
Business Requirements	...	1	1	0	0	1	2	2	1	2	1	2	1	2	2	2	1	1	1	0	3	1	0	3	1	0	0	0	1
Value Stream Planning	...	0	0	0	0	2	1	0	0	1	0	1	1	0	1	1	2	0	0	2	4	1	0	1	0	0	0	1	0
Identification of Improvement Activities	...	0	0	0	0	1	0	0	0	0	0	0	0	0	0	0	4	0	0	0	2	1	0	2	0	0	0	0	0
Definition of Target Conditions	...	0	0	0	0	0	1	0	0	1	0	2	1	1	2	2	4	1	1	0	4	2	0	0	0	0	0	4	1
System-CIP Projects	...	0	0	0	0	0	4	0	0	4	0	4	0	0	4	4	4	0	0	0	3	0	0	3	0	0	0	2	0
Point-CIP Areas	...	0	0	0	0	2	0	0	3	0	2	1	1	3	2	2	0	0	0	1	3	1	0	3	0	0	2	1	1
1.1 System-CIP Execution																													
Target Derivation	...	0	1	0	0	1	0	0	0	0	0	0	0	1	0	0	1	1	0	0	4	0	0	0	0	0	0	0	0
System-CIP Cycles	...	0	0	0	0	0	1	0	0	1	0	1	0	0	1	1	1	0	0	0	1	0	0	3	0	0	0	4	0
Improvement Focus	...	0	2	1	1	1	4	0	0	4	2	4	1	0	4	4	1	4	0	0	1	1	0	0	0	0	0	1	0
Leadership Involvement	...	0	0	0	0	4	0	0	0	1	0	4	0	0	0	0	4	0	0	1	4	0	0	1	0	0	0	1	0
VSM-Quality	...	0	0	0	0	0	1	0	0	1	0	1	0	0	1	1	0	0	0	0	4	0	0	1	0	0	0	1	0
Target Achievement	...	0	0	0	0	0	0	0	0	0	0	0	0	0	0	0	0	0	0	0	0	0	0	0	0	0	0	2	0
1.2 Point-CIP Concept																													
Target Condition	...	1	2	1	0	1	1	1	2	1	2	1	1	1	1	1	1	2	0	0	4	2	0	2	0	1	1	0	2
Quick Reaction System	...	0	0	2	0	0	2	0	0	2	0	2	0	0	0	2	4	2	0	0	4	0	0	2	0	0	0	2	4
Regular Communication	...	0	4	0	0	0	1	2	4	3	1	2	2	4	1	1	1	0	0	1	4	1	0	3	0	0	4	3	0
Sustainable Problem Solving	...	4	3	0	0	3	0	0	4	0	3	0	2	4	0	0	4	4	0	0	4	4	0	2	0	2	1	3	4
Process Confirmation	...	0	0	0	2	1	0	0	1	0	2	0	1	2	0	0	1	0	0	0	2	1	0	2	0	0	0	2	0
1.2 Point-CIP Execution																													
KPI-Effect	...	0	0	0	4	4	1	0	1	1	2	1	1	0	2	2	0	0	0	0	1	1	0	2	0	0	0	3	0
Quality of Problem Solving	...	0	4	2	2	1	1	4	2	1	0	1	0	1	1	1	0	1	0	0	4	1	4	2	0	2	0	4	3
Sum of Reached Maturity Level	...	6	17	6	9	22	20	9	18	23	15	27	12	19	22	24	33	16	2	5	56	17	4	32	1	5	8	34	16

Table F.4: Assessment results per warehouse: control group 2011 C1-C28

G Appendix – Control Group KPR

Description	2010 Lean-Index	2011 Lean-Index	Delta	Slope	2010 Average	2011 Average	Jan 10	Feb 10	Mar 10	Apr 10	May 10	Jun 10	Jul 10	Aug 10	Sep 10	Oct 10	Nov 10	Dec 10	
C1	20.000	24.000	4.000	0.015	90.000	93.750	100.000	80.000	80.000	75.000	75.000	85.000	105.000	95.000	100.000	100.000	100.000	85.000	...
C5	18.000	56.000	38.000	-0.028	101.613	95.699	100.000	100.000	96.774	100.000	103.226	100.000	103.226	106.452	103.226	103.226	103.226	100.000	...
C6	16.000	24.000	8.000	0.045	144.792	160.349	100.000	112.500	125.000	141.129	163.306	149.194	160.887	159.677	143.548	147.581	178.226	156.452	...
C9	0.000	0.000	0.000	0.022	103.561	130.743	100.000	108.782	118.095	101.313	92.026	109.351	112.999	16.919	126.359	124.602	130.701	101.587	...
C10	0.000	15.000	15.000	0.001	100.085	101.984	100.000	99.660	105.102	101.020	109.184	93.878	101.361	99.320	92.177	89.796	105.102	104.422	...
C27	17.000	9.000	-8.000	0.008	108.133	111.667	100.000	109.600	120.000	129.600	129.600	100.000	109.600	100.000	109.600	100.000	100.000	89.600	...
C32	8.000	9.000	1.000	0.002	90.741	91.204	100.000	88.889	83.333	94.444	94.444	83.333	83.333	100.000	88.889	88.889	83.333	100.000	...
C34	8.000	20.000	12.000	0.016	104.832	111.069	100.000	97.318	110.217	110.856	108.046	108.301	98.851	102.427	106.258	105.747	110.473	99.489	...
C35	7.000	9.000	2.000	-0.014	79.821	84.676	N/A	N/A	N/A	N/A	N/A	N/A	N/A	N/A	N/A	100.000	84.115	55.348	...
C37	14.000	23.000	9.000	-0.023	110.140	100.987	100.000	103.318	111.256	111.374	109.597	109.597	113.744	117.891	114.455	110.427	107.938	112.085	...
C38	23.000	15.000	-8.000	0.013	103.419	109.756	100.000	108.362	102.439	105.052	107.317	105.139	101.220	100.261	102.787	98.084	102.003	108.362	...
C39	12.000	27.000	15.000	0.003	110.527	110.059	100.000	107.063	113.644	113.162	111.236	112.841	121.990	115.409	111.236	112.520	110.754	96.469	...
C40	7.000	12.000	5.000	-0.030	100.000	85.366	100.000	100.000	100.000	100.000	100.000	100.000	100.000	100.000	100.000	100.000	100.000	100.000	...
C42	11.000	24.000	13.000	0.034	111.742	122.184	100.000	100.455	105.758	110.455	113.788	115.303	113.788	116.364	115.606	116.364	121.212	111.818	...
C43	11.000	22.000	11.000	-0.035	97.037	83.870	100.000	93.556	94.222	98.444	99.556	106.667	102.667	94.889	91.111	94.000	95.556	93.778	...
C44	11.000	33.000	22.000	0.030	95.897	108.708	100.000	103.311	96.774	97.284	91.426	97.878	98.557	77.844	98.896	98.472	98.896	91.426	...
C49	12.000	17.000	5.000	0.030	132.292	138.356	100.000	94.444	123.611	125.000	158.333	148.611	133.333	140.278	145.833	144.444	151.389	122.222	...
C54	27.000	8.000	-19.000	0.032	91.170	108.319	100.000	93.612	78.109	74.446	80.579	88.245	96.593	90.119	102.129	103.237	102.215	84.753	...
Average	12.333	19.278	6.944	0.005	105.258	108.264	100.000	100.051	103.784	105.211	108.627	106.667	109.244	101.932	108.948	107.633	110.285	100.712	...
Standard Deviation	7.121	12.356																	

Table G.1: Control group KPR development 2010

Description		Jan 11	Feb 11	Mar 11	Apr 11	May 11	Jun 11	Jul 11	Aug 11	Sep 11	Oct 11	Nov 11	Dec 11
C1	…	115.000	90.000	95.000	90.000	75.000	85.000	85.000	90.000	85.000	100.000	100.000	115.000
C5	…	103.226	106.452	106.452	109.677	106.452	112.903	109.677	109.677	106.452	54.839	58.065	64.516
C6	…	152.016	154.839	169.758	178.226	163.306	168.145	180.242	181.855	150.403	138.306	150.806	136.290
C9	…	173.893	184.510	176.090	132.540	160.631	141.146	135.853	7.487	134.598	143.880	113.715	64.571
C10	…	96.599	113.265	109.524	102.381	99.660	102.381	97.959	88.095	100.000	110.884	114.626	88.435
C27	…	88.800	97.600	118.400	132.000	81.600	129.600	100.800	131.200	96.800	94.400	138.400	130.400
C32	…	83.333	88.889	88.889	88.889	94.444	94.444	100.000	100.000	88.889	88.889	83.333	94.444
C34	…	105.619	107.918	106.641	112.261	111.494	110.856	106.130	115.709	115.070	115.198	115.964	109.962
C35	…	92.854	66.818	108.876	73.600	102.777	73.964	97.497	81.111	84.752	85.708	89.804	58.352
C37	…	117.062	111.730	101.896	114.455	109.360	105.687	74.526	78.199	98.815	93.839	102.844	103.436
C38	…	114.111	108.885	109.756	106.272	107.056	111.063	106.882	109.756	108.188	113.937	105.923	115.244
C39	…	113.002	115.249	112.360	105.297	96.308	94.864	100.482	107.384	117.817	118.780	120.225	118.941
C40	…	85.366	85.366	85.366	85.366	85.366	85.366	85.366	85.366	85.366	85.366	85.366	85.366
C42	…	110.758	117.273	116.970	116.364	117.424	126.970	123.788	129.242	128.182	122.273	131.061	125.909
C43	…	99.111	98.667	65.556	80.667	74.667	98.889	96.000	91.778	86.667	68.444	75.333	70.667
C44	…	98.048	105.093	105.688	109.423	105.263	115.110	111.460	92.784	107.301	115.789	117.063	121.477
C49	…	129.306	137.639	102.917	155.972	136.944	147.222	143.611	155.556	140.139	139.028	130.417	141.528
C54	…	103.918	105.622	116.695	132.879	117.547	118.399	95.400	107.325	108.177	108.177	86.882	98.807
Average	…	110.112	110.879	110.935	112.570	108.072	112.334	108.371	103.474	107.923	105.430	106.657	102.408

Table G.2: Control group KPR development 2011

H Appendix – Warehouse Excellence Projects Overview

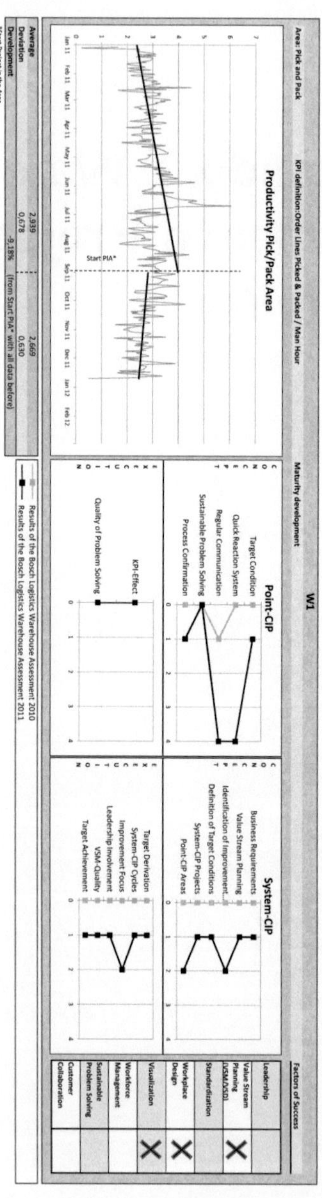

Figure H.1: Project development sheet Warehouse 1

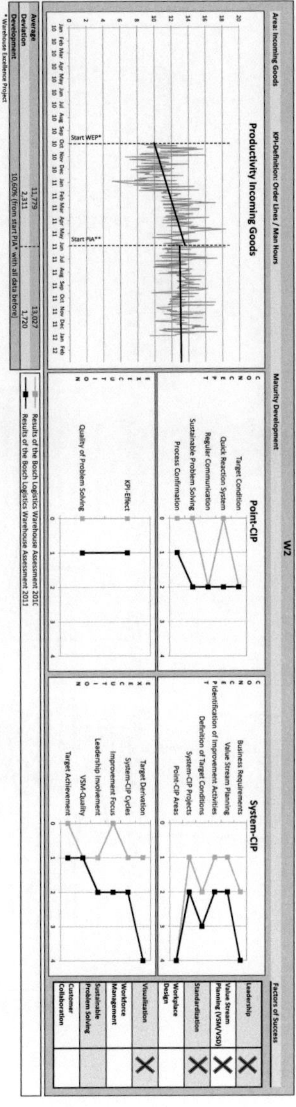

Figure H.2: Project development sheet Warehouse 2

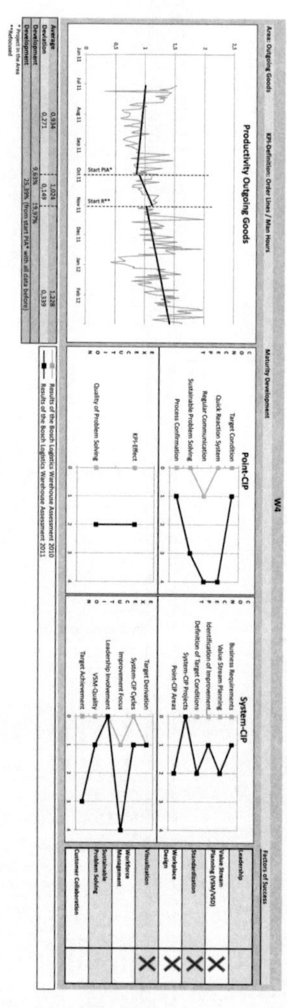

Figure H.3: Project development sheet Warehouse 4

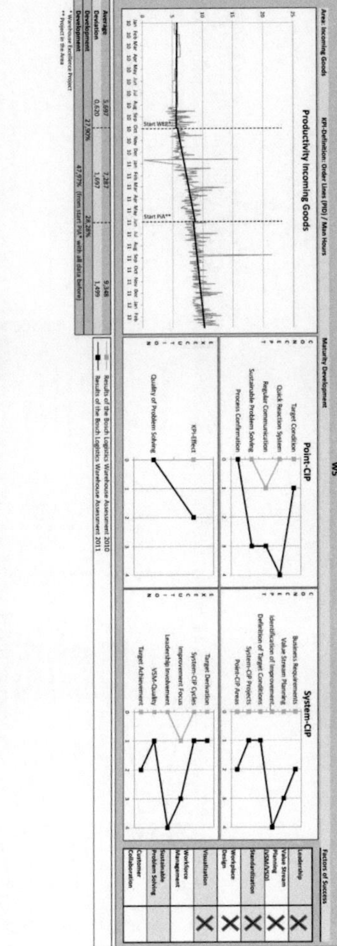

Figure H.4: Project development sheet Warehouse 5

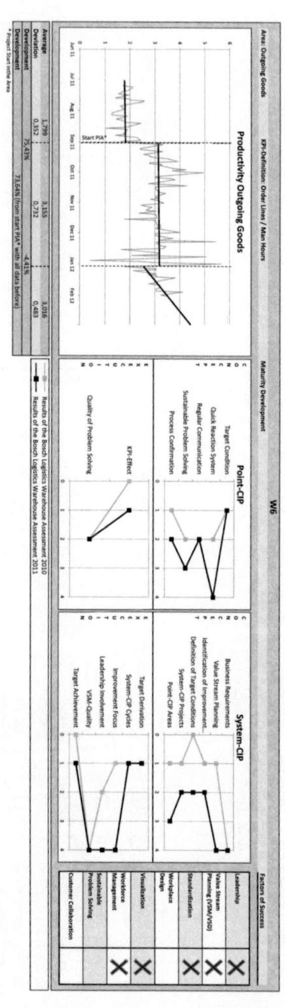

Figure H.5: Project development sheet Warehouse 6

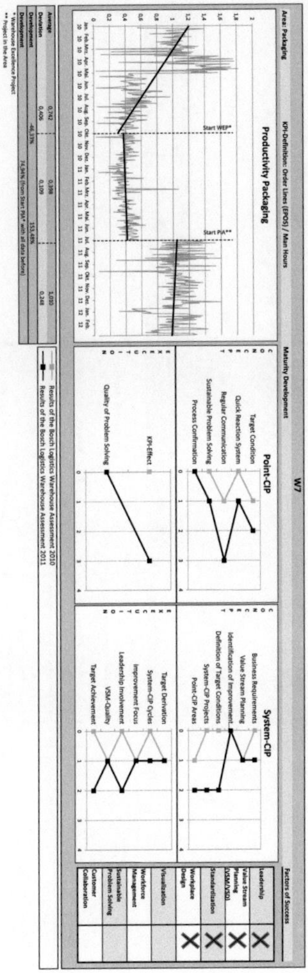

Figure H.6: Project development sheet Warehouse 7

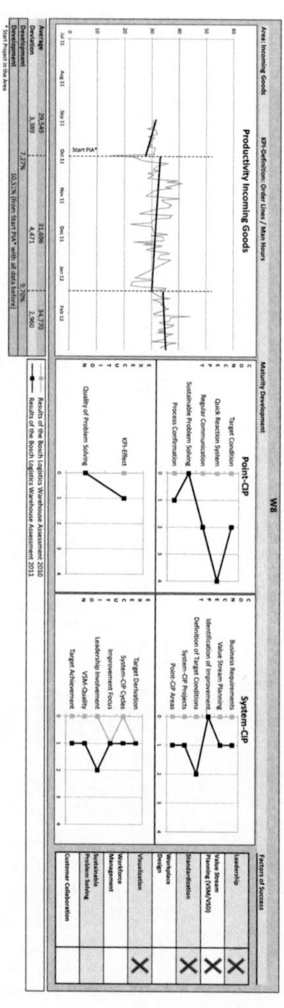

Figure H.7: Project development sheet Warehouse 8

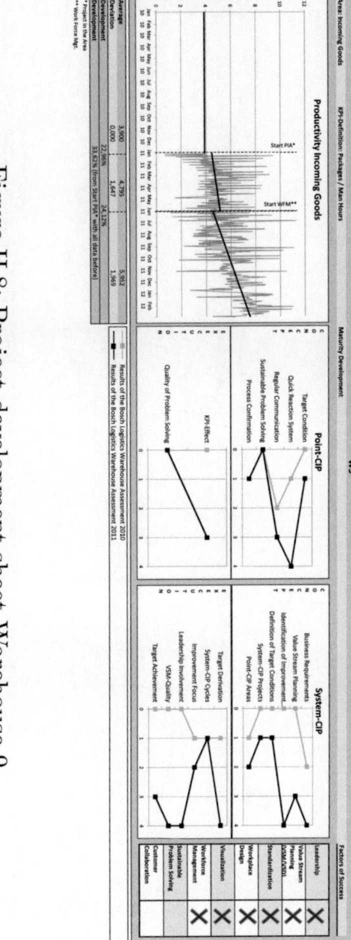

Figure H.8: Project development sheet Warehouse 9

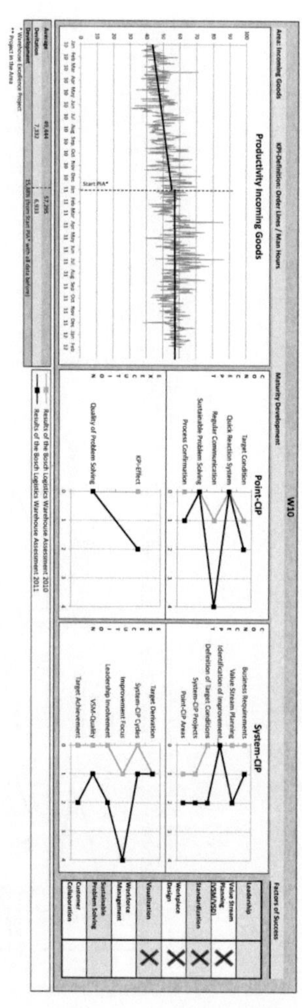

Figure H.9: Project development sheet Warehouse 10

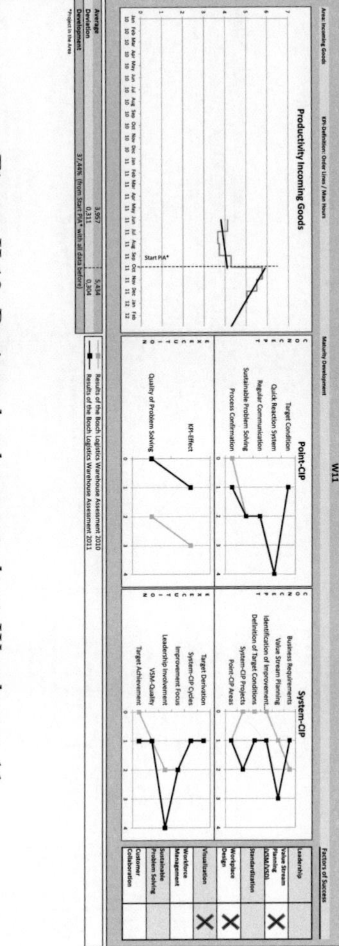

Figure H.10: Project development sheet Warehouse 11

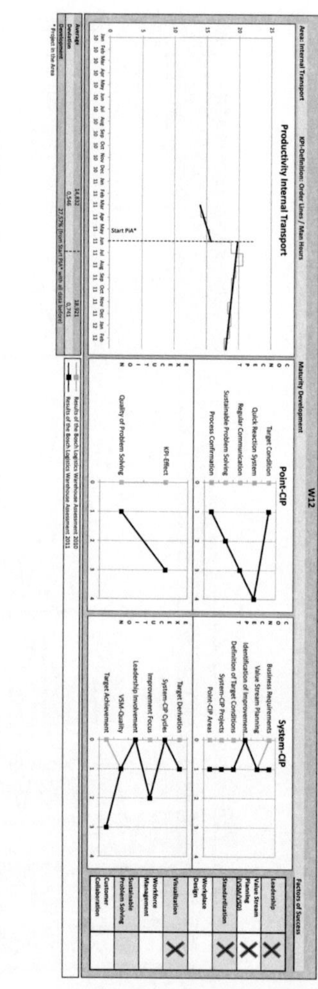

Figure H.11: Project development sheet Warehouse 12

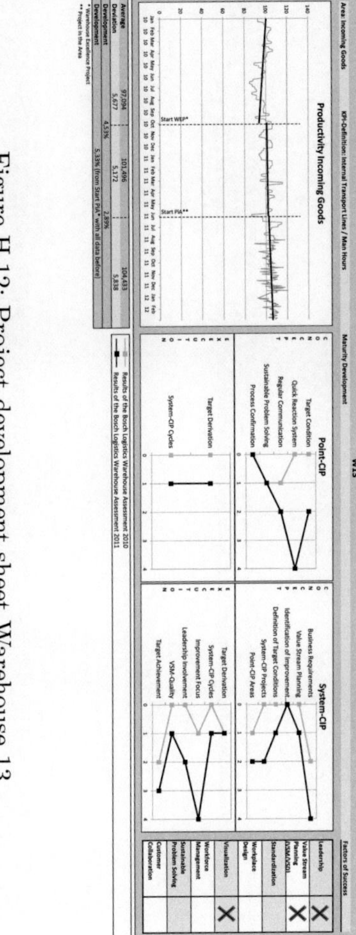

Figure H.12: Project development sheet Warehouse 13

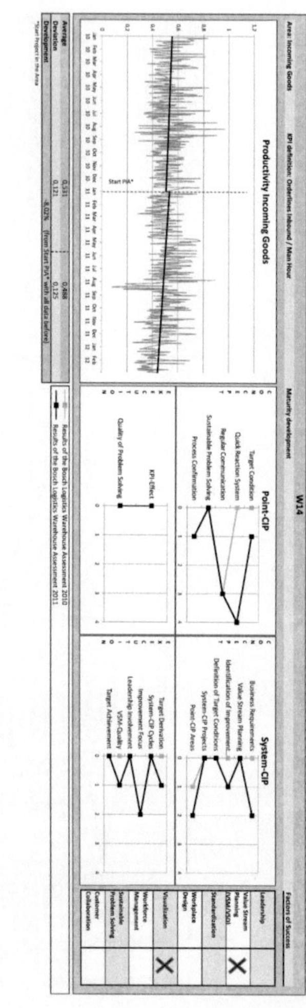

Figure H.13: Project development sheet Warehouse 14

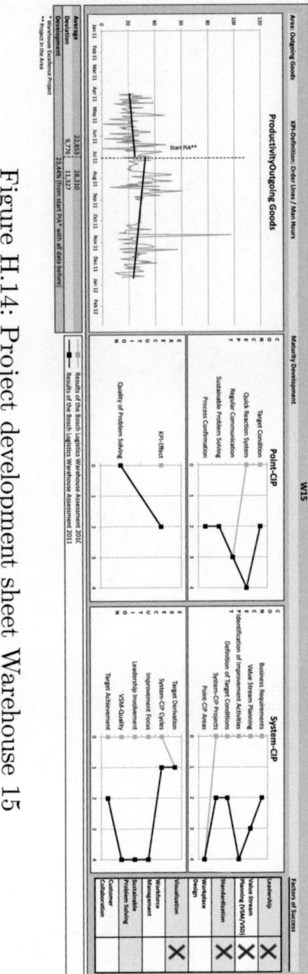

Figure H.14: Project development sheet Warehouse 15

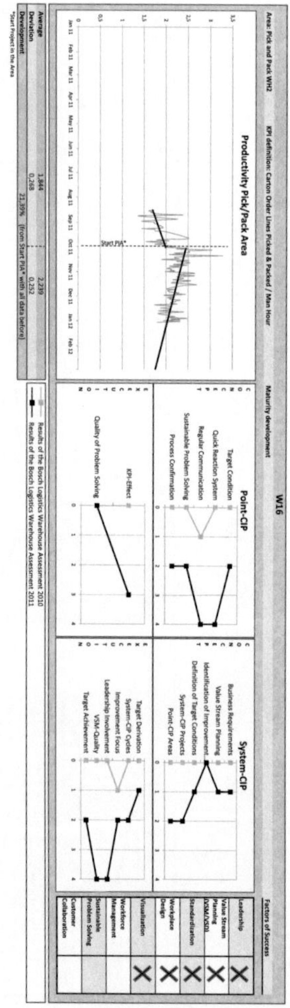

Figure H.15: Project development sheet Warehouse 16

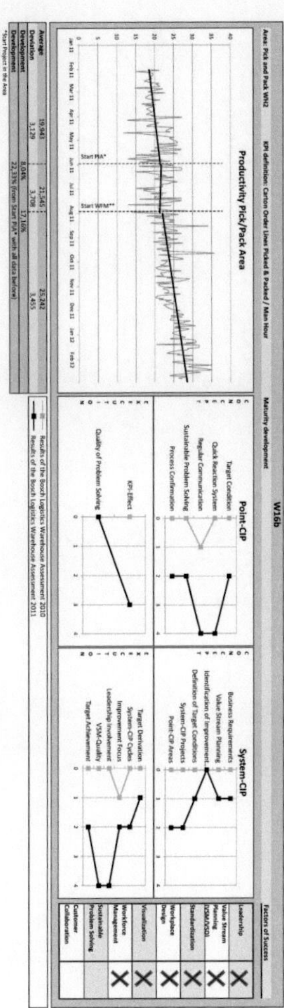

Figure H.16: Project development sheet Warehouse 16b